The Starchild Skull

Genetic Enigma or . . .

Human-Alien Hybrid?

Lloyd Pye

Π

Bell Lap Books Inc.
www.BellLapBooks.com

Praise for *The Starchild Skull*

Lloyd Pye should be identified as a national hero for the work he has done in exploring the mysteries of *The Starchild Skull*. In refusing to abandon this mystery, he has given us all a chance to open our minds in a whole new way. His work is unique in this world, and uniquely important, and I, personally, am deeply grateful to him for his determined effort.

Whitley Streiber, author of *Communion* and *The Grays*

In *The Starchild Skull*, Lloyd Pye has presented the world of science with a genuine mystery. And isn't that what the intellectual quest is ultimately all about? Not just cataloguing the known, the ordinary, but challenging us to enter into the realm of the unknown and the extraordinary.

Michael Cremo, author of *Forbidden Archeology*

Lloyd Pye's *The Starchild Skull* poses an excruciating enigma, which may well lead down a path to the revelation that we are not from this planet. This is a detailed, but exciting, narrative of the search for the truth about human origins. If Lloyd is right, then all of human history will have to be rewritten.

William Birnes, co-author of *The Day After Roswell*

Few men have the perseverance and vision to follow such an incredible odyssey as has Lloyd Pye. Not only a great adventurer, Lloyd adheres to the highest of research and scientific standards on his amazing quest. His dedication to unraveling the genetic mystery of the Starchild skull is inspirational, and sets a standard for us all. This is a journey unique in the annals of human archeology—or in this case, non-human! Bravo to Lloyd for such an unforgettable achievement.

Jeff Rense, Nationally Syndicated Talk Show Host

No amount of research into UFOs and the alien presence prepared me for what I felt in August of 1999 when I held the Starchild skull in my own hands. Lloyd Pye has now thrust this amazing artifact into every reader's hands through *The Starchild Skull*, a gripping saga of his encounters with world class arrogance and closed scientific minds. I know of no other colleague who would have put his life on hold to pursue such a daunting task. Lloyd has laid bare his own emotions and frustrations as he struggled to find the scientific proof needed to establish the origins of this one-of-a-kind skull.

Don Waldrop, Director, Mutual UFO Network (MUFON)
Los Angeles Chapter (1991-2002)

These days rollercoasters come in all shapes and sizes, and Lloyd Pye's *The Starchild Skull* is a literary thrillride. It completes a wild span of sweeps, loops, dives, and figure-eights throughout. At its topmost peaks, Lloyd's broad vision is magnificent, while at the depths of the lowest dives his personal sacrifice to learn the truth about this ancient skull borders on the tragic. This is a damn good book that should be read by everyone with an interest in alternative knowledge.

Andy Lloyd, author of *Dark Star*

The Starchild Skull

Genetic Enigma or . . .
Human-Alien Hybrid?

For information contact:
Bell Lap Books Inc.
38 S. Blue Angel Pkwy
Suite 210
Pensacola, FL 32506
www.belllapbooks.com

ISBN 10: 0-9793881-4-7

ISBN 13: 978-0-9793881-4-9

LCCN: 2007931782

Printed in the United States of America

DEDICATION

_____

To my mother, Nina Pye, who died giving me birth. After being clinically dead for several minutes, she miraculously came back to life. Now she has lived an additional six decades as a widely beloved wife, mother, caregiver, and supporter of charitable causes locally and nationally.

And to my precious wife, Amy, who has birthed me in a very different way, and with whom, at this very late stage in my life, I have finally managed to become an adult.

ACKNOWLEDGMENTS

Eight years is a long stretch on anybody's calendar, so a full listing of those who have aided the Starchild's progress during that time would require twenty pages. Suffice it to say that nearly all of the major players are featured or mentioned at some point in the text of this book. The many others who assisted the Starchild during its journey, with either technical, financial, or emotional support, know who they are.

I wish I could mention everyone by name, but in my last book I thanked everyone I could think of, 75 people, yet I still failed to mention a dozen who let me know they were hurt by being omitted. Rather than attempt the impossible this time, I will err on the side of caution and thank everyone collectively, sincerely, in the wise words of William Shakespeare:

I can no other answer make but thanks, and thanks, and ever thanks.

I deeply appreciate each and every contribution of help.

CONTENTS

PREFACE

All great truths begin as blasphemies.

—*George Bernard Shaw*

For eight years I have struggled to get a remarkable skull recognized and taken seriously by a scientific community reluctant to grant even a sliver of credibility to anything lacking their official sanction. For them, approaching it objectively is indeed a blasphemy, so their skepticism is prompt and pervasive, not just for the usual excuse of winnowing out the all-too-frequent bogus claims, but to avoid having to challenge their own cherished beliefs. Yet because this skull is so ripe with potential to change history, and so solidly documented by numerous testing procedures, science must be—and must remain—at the forefront until the issue is resolved, one way or the other, up or down, win or lose.

Through the text of this book, dozens of grayscale illustrations are placed as near as possible to where they optimally should appear in the natural flow of events. Each reader is free to view the photos or read their captions out of sequence, but that is not the best way to proceed. The book's impact will be maximized by letting the story—and its subsequent discoveries—unfold at roughly the same pace they occurred.

Another way to enhance the experience of reading this book is to know that the first two chapters are used to set up the rest of the story. Don't be confused by how the book begins in Chapter One (*Backstory*). It takes you where you need to go, and then leads into Chapter Two (*Provenance*), which further clarifies what is needed to understand and appreciate the primary story that begins in Chapter Three (*First Impressions*).

Not every person in the book is identified with their real name. Some requested their real names not be used. Others do not appear in a favorable light, yet were, I feel sure, doing their jobs with honest intentions, to the best of their ability. In most cases I've given those people color names: Dr. White, etc. Sometimes I put their names in quotes: "Jane," while other times I use merely a first name with no last name.

A decade or more of specialized training leaves nearly all such "experts" utterly convinced that "natural" solutions are required for *every* questionable phenomenon. I've come to view their blinkered perception of the world with a reasonable degree of sympathy, enough to avoid naming names when it might be hurtful with no other purpose than to be hurtful.

This book is constructed to be easily accessible to everybody, from teenagers to octogenarians. Therefore, it is not clogged with references, footnotes, or endnotes, the literary thicket often used by academics to prove they *are* academics. Ample information about anything I mention or discuss is readily available in libraries and on the Internet, so cluttering the text with footnotes and references seems redundant.

Lastly, any work of this nature must, of necessity, be largely a product of memory supported by notes jotted during or after meetings, letters, emails, and other communications that relate to the events depicted. In every case, I have labored to be true to the essence of scenes and encounters as they occurred. Any errors of fact that might be found herein are solely my responsibility, and, if brought to my attention at lloyd@lloydpye.com, will be corrected in future editions.

CHAPTER ONE

BACKSTORY

A straight line may be the shortest distance between two points, but it is by no means the most interesting.

—*Dr. Who*

900 YEARS AGO

Her name doesn't matter. We assume she was Raramuri, an ancient tribe of Native Americans known today as the Tarahumara. They lived in northwest Mexico, 100 miles southwest of today's Chihuahua, close to the Copper Canyon, Mexico's version of Arizona's Grand Canyon. The area sprawls across a high desert, where baking heat is common throughout much of any year.

Everyone in her time and place would know what heat did to corpses, which probably made her think about the abandoned mine tunnel. Its ever-present coolness would have been inviting to a desert dweller as a place to spend eternity. She had a corpse to bury, after which she meant to lie down beside it and commit suicide. Choosing her own final resting place, and that of her deceased companion, were not matters she would have taken lightly.

She felt the mine tunnel would be a good choice. When she was a young girl straying far from the village in search of food, she accidentally discovered its entrance, overgrown with brush. It came to be the secret place where, during her childhood, she returned for quiet solace in times of stress. She never spoke about it to anyone, and now, many years later, it would be her final resting place. From there, if the shamans were correct, she would join her deceased family and friends—and be reunited with her beloved.

She crept away well before dawn, so no one in the village would know which direction she took. Along worn paths between arid ravines she went high into the hills surrounding the village. Across both arms she carried the frail corpse, as remarkably light in death as in life. A child of the same size would easily weigh twice as much. This was like carrying a dried gourd instead of a melon. Even so, in the morning murk she broke a heavy sweat, reaching the tunnel just as the top of the sun cleared the horizon.

She paused to fully absorb it, the bright orange orb to the east, the nearly full shimmering moon to the west. It was the last she would see of either.

The tunnel's bramble-shrouded entrance was midway up a ravine wall, which meant a struggle to get up to it, cradling even a light corpse. Breathing hard, she gingerly shoved prickles away until she could finally gaze inside. She hadn't been there since. . . . She couldn't recall. Too many years had passed. She was glad to see it was as she remembered, and, not for the first time that day, tears of grief and remembrance stung her weary eyes.

As always, the overgrowth couldn't stop the barely risen sun from illuminating the main shaft well beyond the entrance. She was able to carry her charge fifty paces before losing too much light to take a secure next step. She lay her burden down and sat beside it to rest, letting her eyes fully adjust to the enveloping dimness. By the time her breathing normalized, she could see as during a half-moon night, more than enough for what she had to do.

Choosing the tunnel had been driven by instinct and faded memory, but once inside, she realized it was ideal for her needs. People from her village would search for her, but they could only locate the opening if they chose the correct ravine, which was one among hundreds, and then noticed her tracks leading up into the brush covering its entrance. That didn't seem likely unless she was incredibly unlucky.

More important for her peace of mind was that scavengers weren't likely to find them, either. They were far enough inside the mine's entrance that the air surrounding them was quite still. No odor of carrion would seep out far enough for even rats to smell it. If neither people nor rats found them, they could be there forever. That knowledge eased her grief and reinforced that she was doing the right thing.

She had a small wooden spade strapped to her back, the one she used when tilling fields of maize. She had done all the jobs of women: tilling and planting, harvesting and grinding, cooking and cleaning . . . and loving, too. Her life had been good—hard, but good. She had a brave husband who died defending their village against marauders from the south. She fought beside him, receiving an axe blow to the left side of her head. She nearly died from that and, after it healed, she was never quite her old self. Everyone said so.

She also lost her oldest son in the battle that took her husband and nearly killed her. Later, her two remaining sons died from fevers. She was alone and lonely when her beloved came into her life, giving her a reason to breathe again, to work hard and to thrive. Now, as before, she was alone and heartbroken, but no longer able to face that grim reality. Besides, what did it matter when she died? Death was inevitable and nothing to fear; she was old enough to be at peace with that. She never expected her life to end by her own hand, but she couldn't imagine anyone ever expecting to die that way.

Expected or not, the obligation she had set for herself was upon her, so she rose, refreshed by her rest, to orient herself for the digging. She chose to align the grave on the tunnel's long axis so that if anyone did manage to locate them, they'd be found lying together, side by side. In life her dearly beloved had possessed the largest part of her heart; in death that should—and she hoped it would—continue uninterrupted.

If anyone found them later, her devotion would be clear.

75 YEARS AGO

Her nickname was Pelo, short for Pelo Loco, the wild hair swirling across her forehead in a cowlick. She was fourteen or fifteen, strong-willed and rebellious, like most American teenagers. This day she found herself hijacked into a dingy Mexican backwater, her parents' home village.

Before Pelo's birth, her parents "emigrated" to the U.S. across the Rio Grande River into dry, dusty west Texas. After years of trying, they finally became legal American citizens, like their five children, so now they could cross the border at any time without fearing deportation. To celebrate their new citizenship, they returned to the village they had left nearly twenty years earlier, bringing their family with them.

To Pelo the village seemed ancient, from another century. Set in a small, arid valley surrounded by weathered, shrub-covered foothills, it was far more primitive than her parents' stories conveyed. Yet she had to endure two weeks in it or face their wrath. She gritted her teeth and plastered on a smile as the ordeal began, a welcoming *acogida* where, among other things, Pelo and her siblings were introduced to several cousins, aunts, and uncles. They struck her as awkward, backward, or ugly—all except Roberto, two years older than she was. Him, she'd try to get to know better.

As the day waned, several village children offered to take the young *Americanos*—equally alien to them—on a guided tour of the village. Pelo and her siblings were all bilingual from birth, so language wasn't a problem. One thing that caught their attention was a copse of tall trees a mile from the village, along a ravine that twisted and turned far into the hills surrounding the valley. Bushes, shrubs, and cacti grew everywhere, all around, but other trees were few, isolated, and stunted. A cluster of large ones was a novelty.

"Those trees," Pelo said to Roberto, batting her eyes the way Clara Bow did in movies. "How did they get so big?"

"*Agua*," he explained. "In the ground, buried, where we cannot see it. The roots go down to drink. When our wells run dry, that is where we must dig for water. Without it our village would have died many times."

"*Interesante*," she muttered, making yet another mental note to never—*never!*—end up living in a place like this.

Later in the tour, Roberto and the others pointed out two caves visible in the sides of hills at some distance. Both had irregular vaulted openings and more-or-less flat bottoms, like half a grapefruit lying on a saucer.

"These you must not go into," Roberto said solemnly. "They are taboo because they are too dangerous."

"We have old mines, too!" a girl piped up. She seemed to be charmed by Pelo's twelve-year-old brother, Juan. "More dangerous than the caves."

"Yes, the roofs fall too easily," Roberto added. "You must not go in them, either, but they are not near. The rule is to stay out; that is best."

"Do you never break rules?" Pelo said, again batting her eyes the way she imagined Clara Bow did it.

"Not this one," Roberto said firmly, as the heads of the other children shook accordingly. "People have disappeared in them. It is easy to get lost or have something fall on you."

"Besides," the young girl added, "if you climb up to one, the ground shows your tracks. Anyone could see them. You would be punished."

"How?" Pelo asked. "What would happen to you?"

The children looked confused, as if they had never considered that. Then Roberto's boyish expression hardened.

"I never want to find out."

A week of slow, stifling days dragged past. Roberto stayed as proper as on the first day. Pelo couldn't coax him to touch her hand, much less hold it. He was afraid of her. Boys in El Paso were afraid of her, too. Now all she wanted was to get away, to be alone for a while without the chatter of people trying to make her feel comfortable in a dried-out rat hole.

Finally, at the end of the week the oldest man in the village, the one called *Narizon* (Big Nose), said it would storm that evening. Storms seldom came to such an arid plain, but when they did, Narizon's aged bones warned him far enough in advance to give every villager a chance to prepare for it. In Narizon's warning, Pelo saw a chance to be alone.

900 YEARS AGO

The mine tunnel's floor was rubble and dirt, tracked in or created by the digging of countless men and no doubt some women, slaves of the "ancient ones," sinister conquerors from the south who, the old ones said, demanded extraction of gold and silver from anywhere it could be found.

She was able to shovel into it to midway up her forearm before hitting the solid rock of the floor. It wouldn't provide a deep grave for her beloved, but when she covered the slight torso and limbs with what was there, it would be enough.

Following village custom, she unwound the cloth swaddling the corpse. Teary-eyed again, she gazed at the misshapen body, unlike her own but in its own way possessing true beauty. She eased it into the grave, mesmerized by the diminutive size, shorter than she was by a head or more, and now seeming even smaller. One last time she checked for outward signs of what might have caused the unexpected death.

Worms? Poisonous insect bite? Undetected disease? She had no way to know. Death came on its own schedule, unless—as would happen in her case—its job was done for it.

She turned the first shovelful into the grave, and soon the body was covered and the dirt mounded over—except for one flourish she was careful to arrange. The malformed right hand and part of the slender forearm remained free of cover. Being careful, she arranged it, considered, adjusted it, and then adjusted it again. Finally, she had it where she wanted it.

Now it was her turn. She wasted no time, hurrying into it so nothing would slow her momentum. Her worst fear was losing her nerve, losing the will to die with her dearly beloved. She felt in her heart that she wanted to do it, that she *could* do it, but she wouldn't be certain until she actually did it.

She removed her rough-spun clothes, tucking them in the same niche in the side wall where she tucked her beloved's swaddling cloth. Strapped to her waist was a deerskin scabbard holding one of her most prized possessions: a flint-blade knife that, many years ago, was a gift from her father to celebrate joining her new husband in his hut.

Its edge had dulled with use, but after forming her plan to cope with the death, she'd honed it sharp again. Now she unstrapped it, slinging the scabbard and belt far up into the tunnel, where she had heaved the shovel when she finished with it. She had always been a neat, tidy person, which now was irrelevant. Yet she kept at it, doggedly, as if to stop now, in the last moments, would negate all that had gone before.

She lay down on the rubble beside her beloved, adjusting her position until she could wrap the exposed right hand around the upper part of her left arm. In life they often went along like that, hand in hand or arm in arm, and she wanted to be certain they carried forth into death the same way.

She smiled at that thought, pleased with herself, feeling ready to take the final step in her life's long journey.

She gripped the old knife in her trembling right hand and stabbed it into the right side of her neck, where the big artery pulsed with the life that would soon drain out of her. It didn't hurt the way she expected it to. She felt it slice through muscle and sinew, heard scraping sounds inside her right ear; then she sliced hard outward, which she followed with a vigorous sawing motion until the blade broke free and brought a *swoosh!* sound as a crimson spray spurted from the wound.

I did it! It's done! Fear hadn't stopped her—she was stronger than her fear. An unexpected calmness came over her.

Quickly, before she became faint, she heaved the knife up and back into the tunnel to join the shovel, scabbard, and strap—tidy to the end. She thought she heard it clatter where it struck, but she couldn't be sure with the sound of her neck wound pumping: *swoosh! . . . swoosh! . . . swoosh. . . .*

Softer now . . . fading . . . her mind reeling deeper and deeper into the blackness surrounding her final repose. She felt the tunnel's coolness seep into her naked skin, fully exposed as she lay next to the fresh grave along her left side. *Swoosh . . . swoosh . . .* even softer now . . . very faint. . . .

Soon, in no more than a few seconds, they would be reunited in the spirit world. She hoped they could find each other there quickly . . . she hoped the sacred connection they shared would remain intact. . . always. . . forever. . . .

75 YEARS AGO

Pelo announced that she wanted to pick berries. It was late summer, early August, before the new school year began in Texas. Desert berries such as hackberries, barberries, and madrone were ripe, and she'd already been out picking with her cousins. Ordinarily, she wouldn't be able to go anywhere without an entourage of village kids swarming at her heels, but they were now as busy as their parents preparing for the heavy rain Narizon was certain would come in the night. In places built mostly of stone daubed with mud, places that saw minimal yearly rainfall, a pending severe storm was reason enough to batten down everything to minimize damage.

"Don't go far or be gone long," her mother warned, realizing it might be best to allow her headstrong daughter off the leash for a while. Everyone could see she was miserable under the restrictions of village life. "And put on your long pants. Snakes come out when storms come. Be careful!"

The only available basket was a large, unwieldy one, but Pelo didn't mind. It was light, and lugging it was worth getting away from the stifling rigidity sucking her spirit as dry as the desert surrounding her. She walked fast, anxious to get away, until she reached the copse of trees she noticed the first day. No one in the village could see her there, so she took off at a jog, heading up into the hills surrounding the village. She didn't worry about where the path might lead. Those fifty-foot-high trees provided an unmistakable landmark. *Keep them in sight, you can't get lost.*

Choking with pent-up energy, Pelo practically flew up the incline, rapidly putting distance between her and the village, following the path past ravine after ravine, high up into the hills. She barely took notice of her thumping heart, or the flapping of the basket at her side. She focused on the exhilaration of being alone for the first time in seven days, alone and free to do anything. The potential of it was boundless.

Eventually fatigue and sweating caused her to stop. She gazed around, sides heaving, catching her breath. She was surprised at how far she'd come. She was well into the hills now, still able to see the small green ball of trees on the tan carpet below. She started walking while looking around, seeking something to catch her attention. It was all the same, all sand and rock and dusty brush, jumbled and unremarkable, not much different from El Paso.

She decided to wander up a small ravine. She could have chosen any of dozens, but this one seemed to call to her. In there she'd lose sight of the trees, but that wouldn't be a problem. If she only went in, looked around, and went back out, the trees would still be visible when she returned.

Ten minutes in and it began narrowing to its end. She gazed ahead to see what she could see and noticed a roundish hole in the middle of its left flank. A thick, wide pile of detritus fanned out from that hole down to the bottom of the ravine. Dried branches were jumbled below its bottom edge, meaning bushes once flourished there but were now dead and desiccated. *A mine tunnel!*

Her mind seized on the forbidden fruit, knowing she could explore it now because her tracks would be washed out by the rain Narizon predicted.

It was cool, remarkably cool, considering the temperature outside. Even with a storm brewing, the air outside was torrid. Her sweat-soaked clothes clung to her, chilled her, but she didn't care. She'd made it this far and now nothing would stop her. She'd see what the village taboo was all about, see if she could make it . . . *do what?* Cave in on her? Crush her to a bloody pulp? Goose bumps marched along her spine. She loved it. Step . . . step . . . cautious step . . . cautious step.

Whoa! What's that? A vague, indistinct odor came to her, one she couldn't place. *Rotten eggs? Sour milk?*

She took another step and. . . . *Wha—?* A gasp ripped out of her as she sucked in her breath in recognition. *A skeleton!* Lying on its back, up ahead on the mine tunnel floor, was what looked like a complete human skeleton. *Ohmidios!*

Heart thumping, she inched forward to see it better. Like many girls her age in west Texas, skeletons were not new to her. She had seen horses, cows, dogs, cats. If scavengers didn't find and consume a corpse, its bones would bleach white under the relentless sun. However, the only human skeleton she had ever seen before now was the one hung from a standing-rod in her school's biology lab.

Pelo had signed up for biology this coming year, and the main reason was her fascination with that skeleton, the one kids at her school called "Mr. Bones." Now she found herself gazing at a real one, and she crept toward it with trepidation and utter fascination. *A real human skeleton!*

The first thing she noticed was its small size. She thought it could be a child. *Maybe like me!* A gush of adrenalin shot through her. *What killed it?* Heart thumping again, she gazed around furtively. *Something in here?*

She noticed the left upper arm bone. Something was on it. She looked closer in the faint light. *A hand!* No doubt about it, a hand was poked up out of the dirt and rubble beside the skeleton, and was wrapped around its bicep area.

How did that happen? Was it buried alive? Did it try to get out? She leaned closer. *Whoa! Look at that!*

The hand was odd, contorted, missing fingers or something. *What happened to it?* She looked closer still and saw a bit of forearm emerging from the dusty rubble on the floor, which she saw was slightly mounded . . . *like a grave!* That was *it*, she realized—*the mound is a grave!*

Sweet . . . Mother . . . Mary!

Gathering courage, Pelo shoved a finger into the mound piled near the arm to see how it felt. *Not bad, not too stuck together.* She considered digging it up, wondering if she had a right to do that. *Well, why not? The other one is out in the open, isn't it? What's the difference?*

She dropped to her knees and began scraping rubble away from the arm. More bone appeared. Soon the entire arm was visible with parts of the shoulder. The grave was shallow, easy to clear. Tossing caution to the wind, Pelo began digging with furious energy, determined to get it all out in the open, to be as easy to see as the other skeleton.

In minutes it was exposed. She sat on her haunches, struggling to make sense of what was there. All of it was small . . . smaller than the one on the surface, and misshapen all over. *Big head, narrow little face; it has to be a child. But wait! What if it's a monkey? Or an ape?* No, she decided, it didn't look like a monkey or an ape. It looked more human than anything else. She felt certain it was.

But what kind of human has arms like that, and hands, and legs, and feet? And the head! What a strange shape! What kind of head is that?

A deformed one, silly! It's some kind of terrible deformity.

Gazing at both skeletons, she considered what to do next. She'd been gone over an hour, so it was time to start back. If she didn't return in a reasonable time, someone would go look for her, probably someone who could follow the tracks made by her *gringo* shoes with their distinctive heels. She couldn't risk that, not until after the rain came.

For reasons she didn't entirely understand, she felt compelled to take all the bones with her—both skeletons, full and complete. Maybe her school would want them. Maybe they could be hung on rods alongside Mr. Bones. She might even be given a reward for them! *Yes! That could happen!*

She hurried back to the mouth of the tunnel, where she had left the basket. She brought it in and knelt above the heads of both skeletons so the light from the opening could illuminate them. She lifted the first skull, the normal one, and it felt heavy, the way she imagined a skull ought to feel.

The deformed one was next, and it surprised her. Though its body was shorter by several inches, its head was the same size as the other; but in her hands it felt half as heavy. Then she found that every other bone they shared was equally disparate in weight—skulls, arms, spines, hips, legs, feet.

Adding bones to the basket, she realized she never knew a body had so many. By the time she finished, she was afraid she might have missed one of the small ones—a neck bone, a foot bone, a finger bone. If she did, she knew she'd be too embarrassed to try to present them to her school.

She went back at it, sifting the rubble once more, convincing herself she had every last bone there. Nothing was left behind. She was sure she had them all.

She rose to her feet, brushed off as much dust and dirt as she could, then left, eager to return with her prizes. But she still had a major problem. She had broken the village taboo against going into caves or tunnels.

Walking from the tunnel mouth to the landmark copse of trees, she tried to imagine a way to justify or explain going inside it. *What about just confessing?* That might win sympathy, if not admiration. But if she confessed, would she be allowed to keep the skeletons? They were hers now—*finders keepers!* Yet they had to come from the same village, which was hundreds of years old. It was so old, they told her, they didn't know *how* old—just ancient. The skeletons must be of villagers who died . . . *how?* The ceiling didn't fall in on them, that was obvious. But however they died, she had to assume she wouldn't be allowed to keep them, no matter what.

Maybe I can hide them and sneak them home!

That option didn't last long. *Where?* Everything she had with her was in a duffle bag, surplus from the World War. *Both skeletons together would take up half of that—and the sounds they'd make!* If it was anything like the incessant clatter coming from the basket as she walked, she had no chance to get them past her nosy siblings, much less her parents.

Sensing defeat, she reached the copse of trees. It was now or never. She had to make a decision, but she couldn't.

I need more time!

Frantically, she looked for a place to hide the bones. She scrambled down the ravine, seeking a place large enough to hold them secure, yet obscured. She couldn't find anything that seemed suitable. In desperation, she looked up and inspiration struck. A section of the roots of one large tree had somehow been cleared of everything around them.

Pelo didn't bother wondering what might have created that clearing. She knew only a hollowed-out pocket was above her head and she could reach up and stuff all the bones inside it. They'd be safe and dry, protected from the coming rain until she could decide what to do with them.

Pelo explained her lack of berries and dirty clothes by saying she lounged under the copse of trees for a while and fell asleep. By then dark clouds hung heavy in the west, toward the Pacific, so everyone was simply glad she was back, safe and sound, and nobody questioned her beyond that.

The rains were monumental, like nothing Pelo or any member of her family, much less the villagers, had ever experienced. For most of two days the heavens poured, turning all low, flat land into sticky mud, while carving countless new gullies. Rivulets tumbled down to intersect with ravines, and ravines eventually filled with torrents. Everyone was thankful for Narizon's warning, which had allowed them to prepare well—even for this disaster.

On the third day the sun came out, bright as usual, and the village stirred back to life. The men went to inspect their fields and found no crops were salvageable. The winter stockpiles would have to be purchased. Pelo's father wrote a check for $1000 U.S., enough to buy the corn and wheat and millet that would be needed, though he'd have to borrow it and pay it back in a year of driving his bus overtime. That was a small price for him to be able to relieve the anxiety of those who now looked up to him and his wife for having the courage—and good fortune—to succeed in *El Norte.*

The day after the rains stopped, while everyone in the village focused on repair and recovery, Pelo slipped away to try to retrieve both skeletons. She still didn't know how she could smuggle them past her siblings, past her parents, past U.S. Customs, and into her school, but she was committed to trying. It had become an obsession beyond reason for her. She *had* to try to make it happen.

At the copse of trees she received the shock of her young life. Not one bone was where she left it—*nothing!* The area had been scoured clean by the torrents pouring into the ravine from the hills above. Even now, the stream tumbling through was several feet deep. Its peak height, she saw, was a few feet higher because the water had widened the banks, undercutting four more trees lining it. Her tree, the one sheltering her skeletons, was so undercut it fell across the ravine, its trunk stripped of bark by debris surging past.

"Ohhhhhh, noooooooo!" she wailed. *What can I do?*

Nothing. That was the simple answer. They were gone and that was that, so she turned to walk back toward the village, following the edge of the ravine in the hope that somehow, some way, at least one of the bones might have tumbled where she could see it. If she had nothing more than that, only one small memento of her secret discovery, she felt she could be satisfied.

One small piece . . . please, God . . . just one.

A half mile from the copse, midway between it and the village, the ravine took a turn to the right. Growing on the near bank in the curve of the turn were four greasewood bushes with limp, broken limbs a foot up from their bases, though higher limbs had withstood the torrent. *Ohmidios!* Something was tangled in a couple of the mid-level limbs! *A skull?* Pelo moved closer, hardly daring to trust her eyes. *Yes! A skull!* Not a toe bone, not a finger, a whole skull! She felt like yelping with joy but couldn't risk any villagers hearing her. *Thank you, God! I'll take good care of it!*

She kept silent as she crept to the greasewood bushes, unfolding the gunny sack she brought to carry the anticipated booty. The second of the four held her prize, which up close she could see was the "normal" skull, the one lying on the mine floor's surface. Its soaking tumble through the ravine hadn't hurt it badly. Its lower jaw was missing, but that didn't surprise her because when she loaded them into the basket she noticed the lower jaws weren't connected to the skulls. Also, the right cheekbone was broken off and missing, but the left cheek and the upper jaw were intact.

Gazing at it, her first impression held. She was amazed by how good its teeth were. Stained quite a bit, but free of cavities and well-aligned, though the ones in the back were worn flat, like a horse's teeth. Pelo already had five fillings and, with a taste for sweets, she was sure she'd have many more. *What a lucky person to have such fine teeth!*

She put it in the gunny sack, then turned to look through the jumble of limbs in the rest of the bushes. Two minutes later, in the backside of the last bush, she found the other skull. *Hallelujah!* It hadn't fared as well. The entire upper jaw was torn off, along with most of the bridge of the nose, and both cheekbones were broken at the nubs where they connected to the skull. Now it looked even stranger than before.

Suddenly, nearer to the ground, she noticed something thumb-sized wedged between two finger-sized limbs. It was the right half of the upper jaw, somehow separated from the other half. Two rear teeth were still there, but the three in front, the smallest ones, were missing, lost in the half-mile tumble until the greasewood sieved them from the torrent.

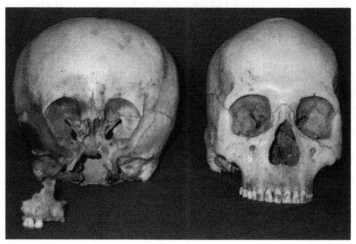

Skulls and maxilla fragment recovered after the flood in Mexico, circa 1930.

The whole lower face was so much smaller than the normal skull's, it seemed a wonder that half of it remained in her hand. She turned it over to look in the front tooth holes. There, she saw beveled edges of new teeth tucked in at the bottoms. No doubt about it, new teeth coming in.

It was *a child! I knew it!*

Pelo put the skulls and the piece of jaw into the gunny sack, wrapping each so they didn't touch. Their bumping together made a distinct sound she couldn't afford to have anyone notice when she returned and went about transferring them from the sack to her duffel bag. Two whole skeletons she never could have managed, but two skulls. . . .

I'll find *a way to manage that!*

CHAPTER TWO

PROVENANCE

The idea that things must have a beginning is really due to the poverty of our imaginations.
—*Bertrand Russell*

Even if beginnings are made necessary by a lack of imagination, they remain important to us all, but especially when an object under consideration could prove to be the most important relic in world history. Such is the case with the genuine bone skull known as the *Starchild*. In these pages you will come to understand its staggering potential, and learn why it might answer the most loaded question humans can ever pose: *Are we alone in the universe?*

A considerable amount of scientific testing has already been done on this remarkable skull, providing enough reliable data to strongly suggest it doesn't resemble the skull of any of the estimated 10,000,000,000 (ten billion) humans who have ever lived on Earth. It seems truly one of a kind.

We still don't know as much as we can know, and one day will know, about the Starchild. In eight years, though, we've learned a great deal about it, and what we know is, for the most part, astonishing. However, we'll begin by focusing on what we don't know—its background story, or *provenance*.

In scientific terms, provenance comprises all that is known

about the discovery and subsequent progress of an artifact or relic, from its original position (*in situ*) through the hands of every person who deals with it. A typical museum artifact or relic will have a complete provenance record, but certain ones do not. Those include countless pieces with archeological value that have been obtained through private collectors, and even from the black market.

While a provenance with a meticulous chain of custody is preferable, it is not the sole criterion, or even necessary, to authenticate an artifact or relic. The obvious reality of its existence is usually adequate. So it is with the Starchild, whose existence is undeniable. However, its provenance is incomplete enough to start its story off on very awkward footing.

Chapter One was an imagined "re-creation" based on what I know, or think I know, about how those two skeletons came to be where they were found. It was based on the best information I have, the few verifiable facts I'm aware of, and the most reasonable assumptions I could make based on what I understand, though all of it could have been quite different.

The native woman (we know it was a woman) could have killed the Starchild (we don't know its sex or age at death) before burying it. Or it could have been killed by someone else, though we assume (or prefer to believe) it died of a natural cause. She could have buried it first, then come back at some time—even years later—to lie down beside it and die, not prematurely by her own hand, but by the end phase of some terminal disease. Possibilities abound.

There is also this: at some point in the woman's life she suffered a serious concussion. The thick bone of her skull's left parietal area has a spidery web of fracture lines around a small central point of sharp impact, much like a rock ding on a car's windshield. Though healed, the injury could have made her crazy—maybe crazy enough to kill the Starchild and take her own life. Here, too, possibilities abound.

My goal was to create a fictional scenario that conformed as closely as possible to the facts as I understood them, and those facts, while straightforward, are woefully incomplete.

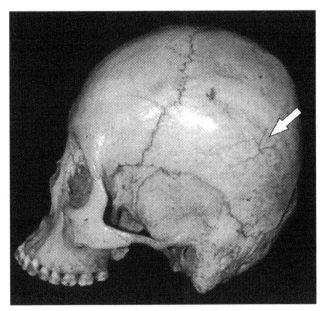

Concussion cracks on the skull found with the Starchild (SFWS).

A woman from El Paso, Texas, found the skulls when she was in her early teens, sometime around 1930, while visiting her parents' home village in Mexico, 100 miles southwest of Chihuahua. This is all we know about the village's location, so identifying and returning to it at present is impossible.

Near the village were several caves and old mine tunnels. She was told not to go in them because they were rickety and dangerous, but like most teenagers warned to avoid something mysterious and potentially dangerous, she ignored the risks and slipped away to do some exploring.

In a mine tunnel—not a cave—she found two skeletons: one supine on the surface, one buried in a shallow grave beside it with a hand stuck up to wrap around the upper arm bone of the one on the surface. In the discoverer's later life, she described the hand as "misshapen," adding that the rest of the skeleton was equally misshapen. She also said its frame was smaller than the one on the surface.

If the skull's oddness is a reasonable gauge of the body's overall shape, it's easy to see why the discoverer described the entire skeleton as misshapen. Also, given that the size of the native woman's skull is smaller than average, she is presumed to be around five feet tall, a common adult height in Mexico today. So it seems safe to assume the Starchild, described by the woman as "smaller," was about four feet tall.

The discoverer did try to recover both skeletons, hauling them away in a basket and hiding them near large trees in proximity to the village, hoping to gain time to figure out how to handle so many bones. A torrential storm did come and washed them all away. She did go looking and found only the two skulls and jaw piece from the "deformed" one.

From the village she was able to successfully bring her prizes back into the U.S., then she kept them for the rest of her life. Early on she shellacked the outside of both skulls, which acted as an excellent preservative for them. Later, she put them in a cardboard box, where they stayed until a terminal disease took her life in the early 1990s, not long after she had passed her unusual legacy on to others.

Solid evidence appears to corroborate this story. Staining on the rear of the adult skull matches staining on the entire Starchild skull. This strongly supports that the adult skull was resting supine on the soil of the mine tunnel's floor, while the Starchild was indeed buried under it. So it is reasonable to assume that if such an important detail in the story can be validated, the rest has a good chance to be proved accurate.

Now consider the discoverer's claim that she exhumed the skeleton with her bare hands. Normally, soil compaction from rain makes that impossible. She'd need a trowel or a small shovel. But in a rainproof tunnel, compaction wouldn't occur. The soil covering the body would be dampened only once, by the dissolution of flesh, blood, and fat by the parasites and bacteria all creatures carry in their bodies while they live. This is what causes decomposition of a body after death.

In non-arid conditions bone, too, would be consumed by bacteria, molds, and fungus. Nature cleans its plate. But in a tunnel with no wind, rain, or airborne spores to provide consumption, bones remain intact. They will desiccate, but they won't dissolve away. So it was in this case, where a pair of well-preserved skeletons were found—one exposed only to torpid ambient air, the other buried in non-compacted soil.

To confirm her story, the discoverer would have to be interviewed, which is ruled out by her untimely death. However, she had a husband, children, friends, and an extended family in Mexico, all of whom know far more about her than what we know now. Unfortunately, here's where things get sticky.

She married a man who worked for a U.S. Government agency, so he was never comfortable with a pair of "smuggled" skulls in their mutual possession, though nothing was legally wrong with it because they were brought in so long ago. Then, in the early 1990s, she learned she was terminally ill. Knowing her husband was not comfortable with the skulls, she decided to secure a new home for them, as one might arrange for pets to be relocated in similar circumstances.

She asked a good friend if he and his wife would accept them. The friend agreed, but he also worked for a U.S. Government agency (a different one), and he was as leery as the first fellow about keeping a pair of human skulls. Again, there was nothing intrinsically wrong in having them, but apparently both men felt that, considering their careers, keeping them provided no benefit and created an unnecessary risk.

The second couple kept the skulls five years, until 1998, when they decided to find somebody with fewer misgivings about owning them. They asked two younger friends, coincidentally named Young—Ray and Melanie—to take them, and they agreed. Melanie had an extensive medical background that left her with no qualms about keeping human skulls. However, when the Youngs saw the weird skull's *degree* of weirdness, they were impressed by how closely it followed the outline of the heads of so-called "Grey" aliens.

Greys are the most commonly portrayed alien form, with small, slight bodies, slender limbs, thin necks, teardrop-shaped heads, small lower faces, and haunting, black, ovoid eyes. We've all seen depictions of them; they're often seen in major movies, TV shows and documentaries, magazines, and books—most notably the startling cover of *Communion*, in which Whitley Strieber famously revealed his ongoing interactions with a group of them over an extended period.

Typical depiction of a "Grey" alien similar to the one seen on the cover of Whitley Strieber's famous book, "Communion". (Image courtesy of artist Rob Roy Menzies.)

Ray and Melanie were members of El Paso's Mutual UFO Network (MUFON) chapter, so they knew how a Grey's skull might look. They told the couple handing over the skulls that they thought the weird one was odd enough to warrant expert analysis to determine what it was. Melanie's medical background suggested to her that despite its strange look, it was probably what the discoverer assumed—a natural deformity of some kind. However, because it was so light and its outline so far beyond normal, and because she knew what Greys looked like, she wanted to be as certain as possible about its pedigree. The other couple agreed, but with a caveat.

Because the first owner's husband and the second couple's husband worked for the U.S. Government, they were well aware that discussing the possibility of alien life forms could lead to trouble with their supervisors, and even sanctions for serious breeches of discretion, up to and including pension reduction or punitive revocation. Thus, keeping one's head down was the best course of action, given the government's well-publicized retaliations against inside whistleblowers and any others who flaunted or ignored the "party line" in matters of official or unofficial policy. So they didn't want their names to be revealed under any circumstances—period.

Ray and Melanie agreed to those restrictions, and the deal was done. They became the undisputed owners of two skulls of dubious antiquity. They knew nothing about them except the sparse story about the discovery in a mine tunnel, and six or so decades in a cardboard box stored in a series of garages. To them, though, that didn't matter. What mattered was finding out, beyond doubt, what the unusual skull was—or was not—or, alternatively, what it might *conceivably* be.

That soon exposed the problems with its provenance.

Any gap in the provenance of an artifact or relic (artifacts are man-made, relics are natural) leaves open the chance of a fraud or hoaxing. The same skills and technology brought to world-class art forgery can be brought to "weathering" and "aging" artifacts to make them look ancient when they are, in fact, recent. So, while a gap in a provenance doesn't *prove* mischief, it creates a possibility, which is often enough to eliminate an item from consideration for that reason alone.

Fortunately, there is no chance of the Starchild being a hoax or fraud. The most cursory examination shows it is made entirely of bone and nothing else. Anyone can see it wasn't cobbled together using disparate bones, as was the Piltdown Man hoax that fooled the science community for forty years. In the Starchild skull, each bone element joins perfectly with those alongside it. However, its individual bones have highly unusual shapes not seen in typical humans, which is what makes it what it is—a *kind* of human.

The overriding question is: *What* kind of human?

It would be wonderful for all concerned if the Starchild skull had a perfect provenance; it would be vastly more convenient all the way around. Nonetheless, it is clearly, stunningly *real*—a bone skull of vaguely human outline that requires a full explanation from scientists who usually spare no effort to avoid dealing with things that threaten their status quos. They will eagerly hide behind a shaky provenance if given even the most tenuous opportunity to do so.

Luckily, it won't always be this way. As stated earlier, the woman who found the skulls had friends and family who know more details about its provenance. Among her friends were the couple who gave it to Ray and Melanie Young, so through them her identity should be within reach. It's simply a matter of giving, to those people who knew her well, enough motivation to identify themselves and tell what they know. That's one of several purposes we hope this book will serve.

We also hope to achieve three other goals: (1) to put the Starchild's case before citizens around the world so they can view, study, and judge it on its merits; (2) to motivate mainstream scientists and the institutions they serve to overcome their natural reluctance to be involved in any issue suggesting a wider reality than the one they currently find comfortable. We hope some bolder ones among them will dare to express an interest in evenhandedly conducting any conceivable test that can wring a shred of usable information from the Starchild's bone. Only then can we all know beyond any measure of reasonable doubt precisely what it is—or is not.

These are expensive matters, so the third important goal is: (3) to interest wealthy individuals who may be inclined to bankroll the blizzard of tests that should be performed on any relic of undoubted historical importance. Consider the many millions spent on thoroughly testing Otzi, the Neolithic hunter found in September, 1991, frozen under a retreating glacier along the Austrian-Italian border. No effort has been spared determining the minutest details about Otzi, which is certainly warranted in a case as unique as his.

Despite Otzi's obvious value to science, he tells us only about our fairly recent history (5,000 years ago). The Starchild skull can do vastly more: it can provide a yardstick for measuring humanity's place in the cosmos. With so much at stake, it deserves no less sparing of effort to determine absolutely everything that can be learned about it. When testing is complete, everyone involved may be able to proudly declare they were a part—however large or small—of making history on a scale few are ever lucky enough to achieve.

With that in mind, the Starchild's story can begin at its second beginning. . . .

CHAPTER THREE

FIRST IMPRESSIONS

The most exciting phrase in science, the one that heralds new discoveries, is not 'Eureka!' but 'That's funny. . . .'
—*Isaac Asimov*

Ray Young is an electrical engineer for El Paso's Power & Light Company. He's tall, solidly built, and matter-of-fact. His wife Melanie is a massage therapist with her own spa. She's as solid as Ray on a smaller scale, with a similarly straightforward manner. They have two dogs, no children living at home, and a hot tub they love to utilize at their house in El Paso's seared suburbs. They would be extraordinarily ordinary if not for being the "parents" of the two "kids" (as Melanie calls them) at the heart of this book.

She calls them kids because, prior to establishing her spa, she was a registered nurse working in a specialty unit for neonatal intensive care. She studied and worked with every manner of human deformity in infancy and beyond, which—more than any other reason—was why she and Ray were asked to take the orphan skulls. The couple who had them for five years felt Melanie could be comfortable with the one they assumed was deformed.

The two skulls in the cardboard box were given to them in October, 1998. Melanie's wide range of professional training provided her with enough knowledge of human physiology to know that the weird skull didn't fall within the boundaries of typical deformity. Also, her association with MUFON (the Mutual UFO Network) gave her enough knowledge about UFOs, and UFOlogy in general, to make a valid connection between the shape and outline of that skull and the alien beings known as Greys. She was convinced that several types of aliens were real, and Greys were one of those types. From her doubly insightful perspective, the weird skull might somehow, someway, be one of theirs.

Starchild skull owners, Ray and Melanie Young of El Paso, Texas.

Like any rational person, Melanie knew her view was extreme, but it was well-balanced by her medical training that told her in all likelihood the skull was indeed some kind of bizarre natural deformity. Nonetheless, until she could be absolutely certain of it, she'd leave her mental door ajar. For his part, Ray Young shared Melanie's feelings on both ends of the scale, so he suggested they should consult with Dan Alegro about their new quandary.

Dan Alegro was a regular attendee at El Paso's MUFON meetings, which they often attended. A manager for a large meat-packing firm, Dan—like most MUFON members—had eyewitnessed a UFO and was certain they were real. When it came to UFOlogy, Dan Alegro was widely read, studious, and well-informed. In MUFON, his opinions were highly regarded, so he was an ideal choice to consult about the weird skull.

Ray and Melanie called to ask to meet with him at a convenient time and place. He suggested they meet after the November MUFON meeting, which they all planned to attend. Dan didn't ask what the rendezvous would be about and Ray and Melanie didn't offer any explanation. That was how it often was in UFO circles. Phones were not necessarily disinterested parties, so sensitive matters were seldom discussed.

During the monthly meeting, Dan was his usual self, chatting and charming his way around the group. Afterward, he and Ray and Melanie moved out to the parking lot to end up standing behind their car. It looked like a routine after-meeting discussion until all of the other attendees were gone. Then Ray opened the trunk, lifted out the cardboard box, and opened it, revealing the skulls inside. Each was nestled in a bed of foam padding crafted by the original owner to keep them from bumping together.

With Ray towering over him in the dim light, Dan looked up from the box, his brow crinkled beneath a snow-white crew cut. "Is this a joke?"

"Dead serious," Ray replied. "No pun intended."

"Are they what I think they are?"

"One is," Melanie said. "It's from a small-statured adult human, five feet or so. But the other one—it's pretty weird, even for me. I've never seen anything like it."

Ray lifted the adult skull for Dan to heft. He gave it a quick once-over. "The back of its head is flat."

"Cradleboarding," Melanie said. "I looked it up. It's caused when a woman straps a baby to a board—especially its head, for safety—so she can carry it around on her back."

Dan nodded. "That was common with natives years ago, and even occasionally today."

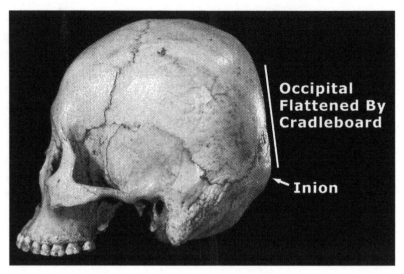

Profile view of the skull found with the Starchild skull (SFWS).

He then turned to take the weird one from Ray in his other hand. He gripped it and his eyes snapped wide. "Good Lord!" he muttered. "This feels like it weighs half as much!"

"Almost exactly half," Melanie said. "We weighed them."

"A little over a pound for the weird one," Ray said, taking back the adult. "This is almost two and a quarter."

Dan studied the back of the weird skull, then made a sweeping gesture around it with his free hand. "All this flattening back here . . . is that cradleboarding, too?"

"Not sure," Melanie replied. "I can't find any references for such an extensive area, so it's probably something else. But the weight is the main problem for me. I *am* sure there is *no* way it should be so light. Not from what I know about human bodies—even if it belonged to a child."

"You think it was a child?"

Melanie lifted the small piece of detached jaw. "This is its upper right maxilla." She held it against her upper right lip. "It fits like this." She handed it to him. "See how small and flat it is? The size of a newborn, yet it had teeth. Now look inside the three empty front sockets. See anything in there?"

Dan's head shook. "Too dark out here."

"New teeth at the tops of the holes," Ray said. "It's a kid."

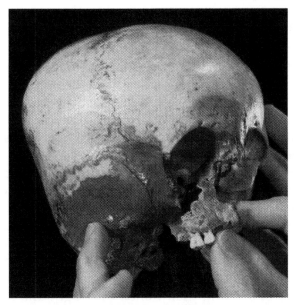

Three-quarter view of the Starchild skull (SC) shows the fragment of its maxilla held in an approximate position of where it would be in the face.

Dan handed the maxilla back, then resumed with the skull, twisting and turning it in the lights of the parking lot, more than bright enough to see the outline of its shape and contours. "It sure does look like a Grey," he concluded. "If the missing face was as small as the jaw piece, it's a close fit."

"Not the eyes," Melanie put in. "They're weird, but not those spooky eyes of Greys."

"The black teardrops?" Dan shrugged. "Don't base anything on them. Many researchers think they're goggles worn for protection—from light, from motes in the air, or whatever. Underneath, their eyes could be more or less like ours. But from the look of such shallow sockets, I'd guess any eyeballs in them would bulge out a lot more than ours do."

"The alien autopsy film showed eyes like ours," Ray agreed, "but I think I remember them bulging out quite a bit."

In 1995, a film surfaced purporting to show the autopsy of an alien body recovered from a widely known, hotly debated UFO crash outside Roswell, New Mexico, in July of 1947.

Its eyes bulged out a bit from sockets that could have been as astonishingly shallow as those staring blindly into that chilly mid-November evening in El Paso. More importantly, thin black "lenses" had been lifted off each eyeball, showing that the strange all-black "eyes" of Greys might indeed not be their eyes at all, but rather lenses used for eye protection.

Among UFO buffs, debate still raged about the autopsy film's validity or lack thereof, but no one could doubt the reality of what Ray, Melanie, and Dan Alegro held in their hands as Melanie put them back on track.

"The rest of this head is different from the alien in the film. If I remember right, that head was a lot bigger and rounder in back. Of course, that could have been extra fat and muscle."

"We can't judge anything by that film because the jury is still out on it. But most eyewitness testimony says *this* is how a Grey skull should look. It has the right size, the right shape, the right weight. Remember, Colonel Corso said the Roswell bodies were short, and their bones were light and strong."

A year earlier, in 1997, the UFOlogy world was rocked by publication of *The Day After Roswell*, the memoirs of Philip J. Corso, a retired U.S. Army colonel who claimed to be an insider to the aftermath of the UFO crash near Roswell. His testimony about aliens and their technology had been subjected to as much scrutiny as the autopsy film, but as Dan Alegro held the Starchild skull in his hands, Philip Corso's words about the alleged Roswell aliens seemed eerily prescient.

Corso had said they were *about four feet tall,* which jibed with the discoverer's claim that the weird skull's skeleton was the size of a child. Corso also said their bones were *thinner but seemed stronger, as if their atoms are aligned differently for greater tensile strength.* Dan couldn't judge the weird skull's tensile strength, but he had no doubt it was thinner and lighter than the other skull.

"That's why we're talking to you," Melanie said. "You know a heck of a lot more about all this than we do."

Dan turned it over once more, carefully, then handed it back. "This could be extremely important. It needs to be analyzed, inside and out, by specialists with credentials, ones whose opinions are respected by all manner of scientists."

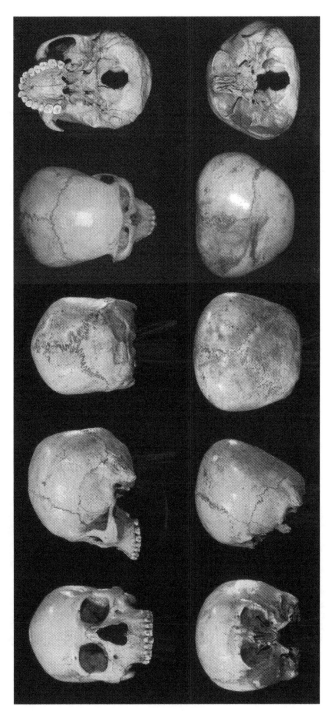

Rotation comparison between SFWS (top) and Starchild (bottom).

"We figured that already," Ray replied. "What we need to know is who we can trust. We were hoping you could help us with that part of it."

Dan shrugged. "I work in meat packing." He focused on Melanie. "What about you? You maintain ties with the medical community. Don't you know anyone who could give you some straight answers about it?"

Melanie considered, then said, "I have a psychiatrist friend who's also an osteopath. We refer patients to each other. She specializes in cranial manipulation for migraines, jaw misalignments, things like that. She may not be a world-class skull expert, but she may know enough to get us going."

"Okay," Dan said, "start with her. See what she says."

Dr. Roberta Fennig didn't mince words. "It's very unusual, Melanie, possibly unique. I'm sure I've never seen anything like this in my practice. You want my advice? Take it to specialists and have it fully evaluated."

"But *you're* a specialist!" Melanie blurted.

"Not for something like this. I don't see enough deformity cases. I'm sure you saw many more of those when you were in neonatal care. How does it strike you?"

"Back in 1930, when it was found, if a newborn's head had one deformity, serious trouble. Two things wrong, very high risk. Three things wrong, almost certain death. This skull has *everything* wrong! All of its parts are abnormal, yet it lived at least a few years, long enough to grow teeth."

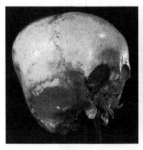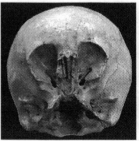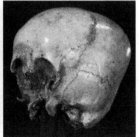

Right quarter, front, and left quarter views of the Starchild Skull.

"In my limited experience," Dr. Fennig said, "deformity tends to be unsightly. This skull, in a way, is beautiful. The degree of symmetry is remarkable. And it's so light!"

"Much too light, much too symmetrical . . . weird."

"I agree. I think you may well have something unique here. If you want to pursue that angle, I can recommend a pair of experts to consult who are far more qualified than me."

"Great!" Melanie replied, so Dr. Fennig wrote two names.

"One is an orthopedic surgeon," she explained, "the other, a pathologist. I could add an EENT (eye, ear, nose and throat) specialist, but I don't know him well. These two I consider friends. I'm sure they'll give you their honest opinions."

Because Roberta Fennig was their friend, the orthopedic surgeon and the pathologist agreed to meet with Ray and Melanie one afternoon at the pathologist's office. The married couple sat, watching and listening, as two bona fide experts tried to determine what they were examining. One would suggest a possible deformity, then the other would offer a better alternative. Soon they drifted into genetic deformities. Lingo and jargon flew like darts around the room, leaving Ray and Melanie duly impressed and hopelessly confused.

"Can you give us a bottom line?" Ray finally asked.

"Most likely a cradleboarded hydrocephalic," the orthopedic surgeon pronounced. "That seems obvious based on its two most distinctive morphological features."

"Notice the adult skull, the flattening at the rear of the head," the pathologist added. "That's cradleboarding. Even today, in certain primitive cultures, all babies are strapped to their mother's back as she goes about her daily work. If the work includes regularly stooping over, the baby's head can't be left to shift freely or its neck will be injured. So it's bound to a board to hold it still, and in a few months the soft bone of its skull fully conforms to the board's flat surface."

"Also," the surgeon added, "hydrocephaly, water on the brain, is a rather common birth defect. Add that to the extensive cradleboarding in the atypical skull, and this is really nothing to get excited about."

"Are you sure?" Ray pressed. "I mean, *absolutely* sure?"

The men looked at each other, and then the pathologist spoke. "Not with only this much to go on. There can always be room for different interpretations."

"However," the orthopedic surgeon added, "you can be confident no reputable specialist will tell you anything other than what we're telling you. It simply *has* to be some kind of natural deformity. Whether it was caused by a known genetic defect or a unique flaw at conception is irrelevant."

"Are there *no* other possibilities?" Melanie asked, trying to coax either man to speculate about the unmentioned issue of alien life forms.

"None we can consider," the surgeon answered. "You see, science has a starting point from which we all work, based on Occam's razor. That's the idea that the simplest solution to a problem is usually the correct one."

"Right," the pathologist chimed in. "So Occam's razor applied to this skull tells us that while it is undoubtedly strange and unusual, it falls within the possible range of natural deformity because that range is, for all practical purposes, infinite. In other words, with deformity *anything* is possible, and given that infinite range, this skull surely fits somewhere."

Based on Dr. Fennig's guardedly positive reaction, the Youngs had worked themselves into a mild frenzy of anticipation that these gentlemen would confirm that their skull was simply *too* unusual to be a typical deformity. Consequently, they left the meeting deflated enough to be ready to tuck the skull box in their garage and be done with it.

What was the point of taking their "kids" to anyone else? Two highly qualified experts had spoken. If they couldn't trust friends of a friend, who *could* they trust?

At the December MUFON meeting they once again discussed their situation with Dan Alegro, recounting for him the "official" opinions they had received. Dan didn't react with the resignation they anticipated.

"They have no way to know if it was a Grey, a Green, or a Blue. Heck, maybe Grey skulls *do* look like cradleboarded hydrocephalics. Maybe those guys saw what they wanted to see, what they expected to see. I doubt it's as cut-and-dried as they said. I think their minds were closed going in."

"Maybe so," Ray agreed, "but they seemed convinced."

"Ha! Don't get me started on ignorant people who think they know what's what because they're in Who's Who."

"Okay," Melanie put in, "what should we do now?"

"Take them outside the doctor club, to someone with no built-in prejudices, no reputational axe to grind. Especially find someone with no commitment to that damned Occam's razor! I'm sick of hearing about it!"

Little wonder. Occam's razor is the weapon skeptics use most often against UFOs. It says that given the lack of hard, can't-be-denied proof of extraterrestrial spacecraft, plus the seeming impossibility of getting from distant stars to Earth in reasonable timeframes, the simplest answer is that UFOs not only don't exist, they *can't* exist. This is based on science's firm belief that no spacecraft could travel beyond the speed of light with absolute security. They reject the possibility that other sentient beings may be able to surpass it because that would upend their gilded seats at the center of our intellectual universe. Such privilege is not easily relinquished.

"If you have someone in mind," Ray said, "we'll be happy to talk to them."

Dan considered a moment. The small but active field of alternative knowledge researchers contained an inviting group of potential candidates.

One might be Whitley Strieber, author of *Communion*, a multimillion selling book with the haunting image of a Grey on the cover. That single painting had solidified the Grey "mug shot" in the public's mind. Whitley's money and influence made him a good choice if he could be persuaded to come to El Paso to investigate the matter, then agree to trouble himself with taking the time to find out for certain what the weird skull really was.

Another might be Linda Moulton Howe, award-winning documentary film producer, ex-beauty queen, and diligent researcher into cattle mutilations, crop circles, testimony by government and military witnesses to the reality of UFO crashes, alien bodies hidden behind maximum secrecy, and many other areas of alternative interest. She, too, could make things happen if she would take time from her busy schedule to visit El Paso and get involved.

Derrel Sims. Dr. Roger Leir. Stanton Friedman. All were possibilities, but their primary interests were not oriented toward relics. The only person Dan Alegro could think of who focused heavily on skulls was . . . "Lloyd Pye."

"*Who*?" Melanie asked. She and Ray knew, or knew of, a wide range of personalities in the field of alternative knowledge. Lloyd Pye wasn't one of them.

"He's a new guy who only started speaking at conferences late last year. He used to be a fiction writer—novels, screenplays, television shows."

"A *fiction* writer?" Ray said edgily. "That's not good."

"It will always work against him," Dan agreed. "But on the other hand, he knows skulls like Colonel Sanders knows chicken. He'd be perfect for this."

"How well do you know him?" Melanie asked.

"Well enough. I met him earlier this year at the Ozark Conference in Arkansas. He was one of the best speakers there. We had lunch together. I liked him."

"What does he speak about?"

"Human origins and what he calls 'hominoids.' That's bigfoot, the abominable snowman, and other creatures like them around the world. He also talks about Zecharia Sitchin's work—all that Sumerian tablet stuff."

Ray sighed heavily. "He writes fiction *and* he believes in bigfoot? This is really not good, Dan."

"Listen, I know how bigfoot *sounds*. But if you saw Lloyd's presentation, all the scientific facts he uses to back up what he says, you'd be as impressed as I am. Everywhere he speaks, he blows audiences away. He's good with words, thinks fast on his feet, and each time he's on the radio with Art Bell, he gets people really cranked up. He's what you're looking for."

Art Bell was the biggest name in alternative knowledge, the keeper of the keys to widespread access. His show went for five hours each weeknight, with replays on the weekend. In 1998 his nightly audience was 5 to 10 million listeners. If Lloyd Pye was a regular guest on Art Bell's radio show, that was like having the Good Housekeeping Seal of Approval.

"What about UFOs and aliens?" Melanie asked, in her no-nonsense way of focusing on the primary consideration. "Does he talk about those, too?"

"Not straight out. He uses Sitchin's Sumerian material to support alien intervention way back in history, when humans appeared out of nowhere on Earth. But he doesn't talk about UFOs or current alien visitation. He avoids anything paranormal or metaphysical, or too far removed from hard science. He leaves all the woo-woo stuff to others."

"Doesn't sound like he's right for this," Melanie pressed.

"Look, he knows skulls because they're a major aspect of his bigfoot research, and he's a great speaker. If your skull is what you think it might be, you'll need someone who can talk to media as well as he can."

Ray and Melanie glanced at each other again, using the silent but effective communication all couples develop over time. A lifted eyebrow, a small tug on a strand of hair. The decision was being made . . . it *was* made.

"Where's he from?" Melanie asked.

"New Orleans."

Ray grimaced. "That's a long haul from here."

"He travels a lot. Maybe he's planning to come this way before long. If so, we can invite him to speak to our MUFON group. But, Lloyd or not, you have to start with someone."

"You're right," Melanie said. "How do we contact him?"

"Leave it to me."

In mid-December, near Christmas, the phone rang in my apartment in Metairie, a suburb of New Orleans. "Hello?"

"Lloyd, Dan Alegro here. How's it going in Cajunland?"

"Fine, Dan! The city always gets decked out for the big holiday. It won't be white, of course. Cold is about as seasonal as things get around here."

"We get white ones now and then, but they're rare."

I let his pause hang without comment so he could get to the point if he wanted to.

"Listen, Lloyd, can you tell me if you think you might be coming to west Texas any time soon? Any trips this way on your calendar after Christmas?"

"In late February I speak at the UFO Congress in Laughlin, Nevada—the big one. El Paso's not on the way, but I can swing by if you think it's important."

"It could be. A couple here has something they want to show you. I'd appreciate it if you'd come by to have a look."

"What is it?"

"That's best left for telling when you get here. But if you'll trust my judgment that it could be important, I'll make it worth your while to swing by."

"I trust you, Dan," and I did. When we met in Arkansas, I was quite impressed by his sincerity and his people skills. "What do you have in mind?"

"You could speak to our MUFON group anytime in February. We're small and can't pay much of a speaker fee, but you'll probably sell your book to quite a few of us."

By then I knew the main roads west like I knew the streets of New Orleans, so I didn't need to check my calendar.

"I have to be in Laughlin late on the nineteenth, so what say we aim for the eighteenth?"

"That'll work. When it's locked in, I'll email the details."

Until a year prior to that call in late 1998, I was a not-very-successful fiction writer. I had published two novels, written for a few television shows, had half-a-dozen screenplays optioned, and generally struggled to secure a well-paying niche in commercial fiction. In frustration, I finally turned my writing skills, such as they were, to an area of my life where I had strong personal interest but dubious commercial potential—alternative knowledge.

During my years of struggle to become a successful fiction writer, my hobby was reading anything I could find about *hominoids*, a term that technically describes the Hominoidea, a superfamily of primates (humans, chimps, gorillas, orangutans, gibbons), as well as their presumed extinct ancestral forms known collectively as *Miocene apes*. I use "hominoid" to refer to upright walking, hair-covered creatures (bigfoot, sasquatch, yeti, etc.) that are not nearly as extinct as commonly believed, but which are dismissed outright by mainstream science and media as UFO/alien type fantasies.

Relatively few people knew of my interest in hominoids. It wasn't the kind of thing to endear me to acquaintances, much less family or friends. It was, instead, a subject that people shied away from in droves because all forms of media saw to it that in "sensitive" areas that might offend scientists or religionists—both of whom were equally thin-skinned and defensive of their dogmas—ignorance of fundamental facts was approved of and promoted as a routine policy.

By early 1997, I had written a book called *Everything You Know Is Wrong: Book One—Human Origins*. Despite many years of disappointments with my fiction writing and screenwriting, I expected this book to be well-received by editors in mainstream publishing houses in New York. It was standard nonfiction, always an easier sell than fiction; it broke much new ground in several areas of alternative research; and I was willing to promote it night and day.

With all that going for it, I expected little trouble getting it published, but I couldn't entice a single literary agent (by then I knew a few) to read it, much less represent it to publishers. Nobody wanted to see a "fringe" book with such an in-your-face title. Two mid-level agents said they'd look at it if I'd change the title, which I wouldn't agree to do because the book's core theme was that virtually everything we think we "know" that has current scientific significance will, in time, be seen in the same light as every other bit of "certain" knowledge in history—as flamingly, stupidly wrong.

I decided to publish it myself, after which came the endless grind of promoting it, out "on the road," facing challenges I never considered before embarking. For all of 1998, I traveled around the U.S. giving slide-backed presentations about my new book (*EYKIW* for short), selling copies at bookstores and alternative gatherings where I spoke. In 1998, I drove 70,000 miles to sell 15,000 books from the trunk of my car, while gaining a reputation as a good new speaker in the field.

Now 1999 loomed as another year of long-haul driving to share my message and sell even more copies of *EYKIW*. It was hard going, the hardest of my life since I quit playing football thirty years earlier, but I was getting it done and seeing positive things happen. I just had to keep them coming.

I would arrange presentations along circuitous routes that focused on at least one major venue and a few smaller ones. On the trip beginning in El Paso, I had six lined up: two in Colorado and one each in Texas, Nevada, Arizona, and Arkansas. This trip's biggie was the second, in Nevada.

The circuit would require a full month to complete. That was as long as I dared to be gone because receiving email away from home was yet to be made readily available. I'd return to find 1,000 to 2,000 emails waiting, which required eight hours a day for three or four days to plow through.

My clothes were hung on a metal bar between two hooks above the rear windows of my 1992 Buick Roadmaster. I'd cram its trunk with boxes of books—20 boxes, 20 books each, 400 total. If I sold one-half, 200, it was a break-even trip. If I sold three-quarters, 300, it was an *outstanding* trip.

Sometimes, I got lucky and sold them all—but that was usually too much to hope for.

Lloyd Pye with his much-loved, well-traveled, somewhat battered 1992 Buick Roadmaster, at the end of another long cross-country journey.

CHAPTER FOUR

MEETING THE KIDS

Discovery consists of seeing what everybody has seen and
thinking what nobody has thought.
                                                   —Albert Szent-Györgyi

Depending on direction of travel, El Paso is the beginning or end of 880 brain-numbing miles of Inter-
state 10 as it threads across south Texas. The city is
sprawled across a bone dry, dusty desert that is as lacking
in appeal to a Louisiana native as grits to an epicure. I came
from year-round green and damp—this was forever tan and
dry. I was always glad to see El Paso fading in my rearview
mirror, so it felt strange to be seeking the Holiday Inn where I
would speak to the local MUFON group that evening.

After several email exchanges with Dan Alegro, we arranged
to meet in the hotel lobby at 5:00 p.m., two hours before my
presentation kicked off. Dan gave no clues about what his un-
named friends intended to show me, nor did I ask. I'd been in
the field of alternative knowledge long enough to know para-
noia was rampant on two fronts: against the government,
which was notorious for using heavy-handed tactics against
anyone claiming serious UFO knowledge; and within the field
itself, where researchers often worked against each other because proprietary information could be so valuable.

I entered the Holiday Inn lobby expecting to be shown something relating to hominoids, in the vein of other items I'd already been shown by other people seeking my opinion. One man had produced a faded photo of a set of bigfoot tracks, taken by a deceased uncle, who never told anyone where he took it. From a muddy ditch came a plaster cast of what might have been part of a huge palm and a base of a thumb—twice the size of a human hand. Or the old gent who showed me a bent piece of oak limb he swore he saw a bigfoot gnawing when he was a boy. Now limb and man were desiccated, so nothing worthwhile could be recovered.

I expected to be shown something like that: interesting, but essentially useless.

Dan Alegro was my age, about fifty, with crew cut white hair offset by a devilish dark goatee. We spotted each other in the lobby, greeted with a hug, and then he introduced me to Ray and Melanie Young. Ray had the size and look of an NFL defensive end, with a sparse hairline that sprouted a ponytail dangling to his shoulders. His left arm cradled what looked like one of my book boxes.

Melanie had the firm handshake of the masseuse she now was. She had an impish smile and was a paragon of polite solicitude, inquiring about the rigors of my trip, how tired it must have made me, and would I be able to get through the evening okay? I liked them both from our first words of greeting. They glowed with the same cheerful, upbeat attitude Dan Alegro exhibited. I could see why they were friends.

We went to the sitting area, where a couch and two armchairs were arranged near an unlit fireplace. People around us could move to and fro all they wanted. However, nobody sat near us and the lobby's foot traffic was light, even at five o'clock, so we settled down to do our business.

Ray and Melanie sat opposite Dan and me, with a small oval coffee table between us. Ray put the box on the table and matter-of-factly said, "We appreciate that you've come out of your way to do us this favor. We have this thing we need some serious help with, and we're hoping you'll find it of enough interest to want to give us that help."

I smiled. "If Dan says it's worth seeing, I believe him."

Melanie reached out and gently pushed the box toward me. "Prepare yourself for a little bit of a shock . . . then open it."

That warning caused me a moment's unease, but then I went ahead and lifted the box's four top flaps. When I did, I could see the outlines of two skulls resting on the box's bottom. That erased my anticipation about this meeting somehow relating to hominoids. Both lacked the heavy brow ridges I had always expected any specimen would have.

"Take out the one on the left," Ray told me. "It's normal."

Indeed it was, a well-preserved cranium, with its upper jaw and teeth intact, but minus its right cheekbone and its lower jaw. The rear of its head showed the unmistakable sign of cradleboarding as an infant—an area the size of my palm pressed as flat as the board to which it had been strapped.

Ray reached to take it from me. "Now the other one."

I did as instructed, feeling overwhelmed by several sensations and thoughts. My first impression, before I could see it clearly, was how amazingly light it was. In my hands, it felt half as heavy as the other skull, which was a good guess.

Ray could read my reaction, so he explained. "It's close to half as heavy . . . almost exact. We weighed them."

As it cleared the top flaps of the box, its bizarre shape stunned me. It had expanded parietal lobes (upper rear) with a distinct crease between them. It was so extensively flat in back, I could hardly believe my eyes.

Speaking of eyes, the blank gaze of its empty sockets sent me reeling on a brief mental jog. For more than forty years my father practiced optometry, so I knew about eyes and eye sockets. These were so wildly unlike what I knew they should be, I struggled to wrap my mind around them. *How could they be like this? So fantastically far from normal?*

Keeping that skull in my right hand, I reached to Ray for the first, holding them side by side over the box between us, ignoring those in the lobby who might have noticed what I held. With Dan in a suit and me in a coat and tie, we could have been FBI agents reviewing a cold case with witnesses. Whatever was imagined about us, no one approached.

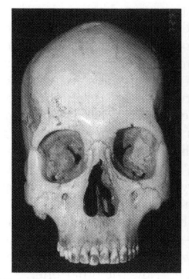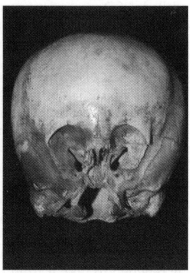

SFWS (left) has normal eye sockets about two inches deep. Starchild (SC), at right, has sockets no more than one inch at their deepest point, which is at the bottom of the orbit rather than in the middle, as is normal.

"What do you think?" Melanie asked, tentatively.

I wasn't thinking, I was drifting, absorbed in a powerful image I couldn't shake. Most times when I looked up into the sky on a clear night, it was just the sky at night. I saw stars, maybe the moon—it was routine, familiar. But every so often I'd look up and *wham!* I'd suddenly grasp the *vastness* of it all, and I'd realize how unspeakably complex it was, and how grossly insignificant I was within that overwhelming context.

That was what I saw in those empty eye sockets—a compact infinity of possibility, of confusion, and of challenge.

"Wheeeewwww," I whistled softly. "This is just amazing! Where did you get it? I mean, get *them* . . . both skulls?"

Human skulls weren't common items to possess, so I listened as Ray and Melanie took turns telling me the curious tale of how these were alleged to be found, and how the box made its way to them. I didn't like the secrecy oath regarding the previous owners' backgrounds, much less the sparseness of the story. It had holes big enough to drive tanks through.

"You look like keeping secrets is a problem," Ray said.

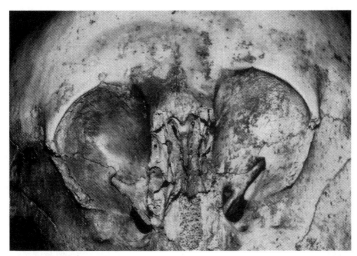

SC's eye sockets are exceptionally shallow and unnaturally symmetrical.

I gazed across the table. "I was in Army Intelligence, so I know how to keep secrets. That doesn't bother me."

Dan smirked at my side, so I added, "Yes, I've heard the oxymoron joke, and I have to admit there's some truth in it."

I turned back to Ray and Melanie. "What *would* bother me, a lot, would be you two not telling me relevant things."

"We told you everything we were told," Melanie insisted.

"It really is all we have," Ray added. "So let me ask you straight up. Do you think you can do anything with it?"

I shrugged. "That's hard to say. I mean, it *is* a strange skull. Anyone can see that. But exactly what it is, I don't have a clue. Probably a deformity of some kind."

"What we need to know," Dan put in, "is *what* kind?"

I shifted my gaze from the skull to glance sideways at Dan. "I know about the skulls that relate to my work, the ones called prehumans. But the limit of my expertise is the details of how they compare with humans. When it comes to deformities like this skull might exhibit, I don't know much."

"Can you take it to anyone who knows?" Melanie asked.

"I'm sure you have specialists like that in El Paso."

Ray shook his head, shifting his ponytail. "We tried a pair. Not helpful. All they came up with was that this—" he hefted the weird skull "—was a cradleboarded hydrocephalic."

"That's not cradleboarding," I replied, pointing to the adult skull's palm-sized, board-flat rear. "*That* is."

"What about hydrocephaly?" Melanie asked.

"If water on the brain caused enough internal pressure to expand the parietals outward, that crease along the sagittal suture, between them, should get pushed out, too."

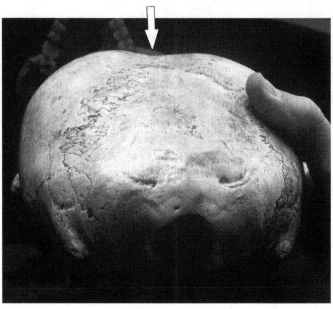

Angle from the rear bottom of SC showing the crease along its sagittal suture. ("Sutures" are the crack-like lines that separate the different plates of bone in all human skulls. The sagittal suture runs along a midline from the top-center to the rear crown of any skull).

Dan winked at his friends. "I told you he knew skulls."

I raised my hand. "Whoa! Not so fast! I don't know *what* this is, other than it's not cradleboarded or hydrocephalic. Anyone with any degree of expertise can see that."

Melanie chortled as Ray growled, "The two specialists we talked to would give you an argument about that!"

Dan stifled a laugh as Melanie pulled us back on track.

"Lloyd, Ray and I need someone to help us find out exactly what this skull really is. That's all we want, to just *know*, for sure; to be absolutely certain of what it truly is."

"And you want me to do that for you?"

"If you're willing," Ray said. "We know you have your life to live, but if you could maybe work this in on the side. . . ."

He left it hanging, so I turned to Dan. "Is there nobody else here in El Paso you can consult? Maybe anthropologists at UTEP?" I knew about the University of Texas at El Paso because they had well-regarded sports teams.

"This is a conservative city in a conservative state," Dan explained. "Finding open minds here will never be easy, no matter how we go about it."

"It's tough anywhere. We live in a conservative country."

"There must be open minds *somewhere*," Melanie insisted.

"It's the law of averages," Ray added. "And you're in a much better position to find them than Melanie or me. We don't look or sound like we know a Halloween skull from a real one. Someone like you, they'll respect."

I shook my head. "No, they *don't* respect me because I'm not part of their system, yet I have the audacity to challenge what they preach. To them, I'm the religious equivalent of a godless heathen. Mostly, they think I'm nuts, so don't kid yourselves. I'll be swimming against a rip tide all the way."

"But you can get to them, can't you?" Melanie pressed. "You can get them to talk to you, to listen to you . . . right?"

"I can get *to* them, sure, but I can't make them talk to me or listen to me. Some will, but most won't, because people like me are threats. Not serious threats, as long as they can keep us trapped in the tabloids, looking like we don't belong on the same field. But we annoy them because we promote ideas that challenge them: ideas they don't understand, don't know about, and don't *want* to know about. So I'm always surprised whenever a mainstreamer is decent to me."

"All it takes is a few," Dan muttered. "The right few."

We came to the crunch when Melanie tired of my waffling. "Will you do it, Lloyd? Will you take the skulls and see if you can find out what the weird one is?"

"I don't have time to consider it carefully enough now." I looked at my watch. "I have to get everything set up for my presentation. Can I have overnight to think about it?"

Ray and Melanie worked it out with only their eyes.

"How about breakfast at the IHOP you drove by coming in?" Ray suggested. "Eight o'clock?"

I nodded as we rose. "Eight at the IHOP, see you then."

International House of Pancakes, IHOP, is one of many symbols of American homogeneity you notice if you spend a lot of time on the road. Waffle House. Cracker Barrel. Wal-Mart. Shell. They make everywhere in the U.S. feel like home. That's one reason I like IHOP. Another is their syrups.

"Do you always put some of each syrup on your pancakes?" Ray asked me, sounding genuinely interested.

"If I'm in a Waffle House, I'll use honey. Honey's better for you. In an IHOP, I go straight for the scrumptuous syrups."

Melanie nodded agreement. "This is my favorite place to eat breakfast. Makes Awful House seem like a dog house."

"Awful House! Ha!" I laughed. "Good one . . . I like that."

As we ate our food, Ray got down to business. "Like I was saying, Dan sure had it right about your lecture. The audience last night was hanging on every word."

I shrugged. "It's compelling material."

"That's why we bought your book!" Melanie exclaimed.

"I assumed you were bribing me," I said in a teasing tone.

"We would if we knew how!" Ray teased back, then he got serious. "We really want you to handle these skulls for us."

"Well, I admit I'm interested in this. Who wouldn't be? But the skulls don't come with a handbook, so if I take them on, all I can promise is to try hard to do my best with them."

"That's all we expect," Melanie said. "If we had any bright ideas about how to get the job done, we'd do it ourselves."

"Okay, then. I'd like a free hand to tack wherever the winds of fate blow me. If you'll grant me that concession, I'll do everything I can to avoid getting grounded on any shoals."

"Hey! Do you sail?" Ray asked. "We love to sail! We go bareboating in the Caribbean every year!"

"I get seasick if I snorkel on water with small waves."

"That's too bad . . . sailing down there is awesome."

The amusement of the moment faded, so I went back to business. "Something else to discuss is the parameters I'll be working under. I want to make sure I don't overreach my responsibility or my obligations to you."

"Whatever you feel you need to do," Ray said, *"just do it!"*

"We'll always back you up," Melanie added.

"What about money? Are you two in this for money?"

They looked at each other, seeming confused, then Ray carefully said, "Noooooo . . . are you? We can't pay much."

"I'm not asking you to pay me. If I do this, it's for the same reason you said—to find out for sure what it is. Besides, there shouldn't be much cost involved in asking some experts to run a few tests on it. Hopefully, for something as unusual as this, they'll do it out of their own pockets."

"That's good to know," Melanie said. "We don't want it to cost you more than the time and trouble to take it around."

"I can afford that, but we need to be clear on another important point. If the skull turns out to be anything other than a total human, you'll get a *lot* of money for it. If that happens, I'd expect a fair reward. Nothing ridiculous, just fair."

Ray grimaced slightly. "How 'fair' are you talking?"

By then I knew them well enough to be comfortable putting our business relationship on the honor system. "You're good people, so I'm not worried. I know you'll be fair."

Both showed relieved expressions. This was clearly an issue they had discussed, and probably fretted over, but didn't know how to broach it with me. Now it was resolved.

"Also," I went on, "if this is the historic relic we think it could be, it has to be treated in a very special way or it will lose its credibility right out of the gate. If I have to hustle it or hawk it or do anything like that, I'm not your man. We can't ever have T-shirts with that image, or bumper stickers, or fright masks, or any of that UFO-alien hokum you see at all the conferences. We have to play it straight as a scientific enquiry, from the start until we have solid evidence in hand."

Both foreheads wrinkled as we finally hit our first snag. Ray said, "You, uhm, have a problem with aliens or UFOs?"

"When I was twenty-eight, living in Palo Alto, I saw a UFO. A bright, cloudless day—I couldn't miss it ripping across the sky. No question, no doubt. So I don't just think they're real, I *know* it, but I don't study them. They've never interested me as much as hominoids and human origins. I can't explain it, that's just the way my mind and my emotions are wired."

"Will you be comfortable working with a skull that might be from an alien?" Melanie asked. "I mean, if it's not something that interests you, can you actually take it seriously?"

"Its potential importance is what I consider paramount, not how I feel about it personally. I have to tell you up front that I assume it's a deformity of some kind. I have nothing against the idea of it being an alien, but I'm going into this to let the chips fall where they may—period."

"How will you get scientists to work with you?" Ray asked.

"By doing everything the way an accredited anthropologist would handle a potentially important fossil find. If I stick to their guidelines, a few of them should be enticed to pitch in and help me. That's why we can't muddy the water with UFO hokum or financial shenanigans. Only certified mainstream scientists can do the kind of testing we need done to figure this thing out, so we can't afford to put them off helping us."

Two vigorous nods. "Sounds good to us. Right, Mel?"

"Absolutely! Whatever you say, Lloyd."

Ray glanced at his watch, signaling he needed to be starting on his way to work. "What's your first step?"

"I'd like to take them to the convention where I'll be speaking this weekend, the UFO Congress in Laughlin, Nevada. Maybe Dan told you about it."

Ray nodded. "He said five hundred people will be there."

"At least. It's the biggest event of its kind in this country. A solid week of speakers, thirty-some-odd. The timing of it, right after us meeting like this, is amazing."

"You won't talk about the skull there, will you?" Melanie asked. "Not before you have an idea of what it might be?"

"No, but I do want to show it privately to some of the cream of the UFO crop that will be there. I'd like their input about its possible relationship to Greys. If their consensus is that it looks like the skull of a Grey, I'll have that much to go on."

Ray leaned forward and dropped his voice a notch to caution me. "You shouldn't tell many people about it. Our government doesn't let things like this, UFO related things, float around unattached. They'll confiscate it or steal it or who knows what, and then we'll never find out what it is."

"I realize some people aren't as discreet as others, so word about it might well leak. But we have to take that risk."

Melanie's concern was obvious. "Why can't you carry it straight to the people, the scientists and such, who can tell you what you need to know about deformity?"

I had to be honest. "You have to understand that I'm not what you'd consider a 'UFO person.' I'm really not very knowledgeable about them. So before I ask any scientists to take this skull seriously, I want some qualified UFO experts to tell me what you two suspect about it isn't dead wrong."

Melanie stared hard at me. "Do *you* think we're wrong?"

I spread both palms outward. "No offense intended, but taken alone, in isolation, your opinions don't sway me."

Ray checked his watch again. "So if you get a few UFO honchos to agree it looks like a Grey, will *that* sway you?"

"With you two and Dan, I'm interested. Add several of them, I'll be impressed."

"How long will it take?" Melanie asked. "From today until when you think we might have a final answer about it."

"Once scientists come to know how incredibly important the skull might be, I suspect they'll be fighting each other to be a part of working with it. Let's say six months at most."

Ray and Melanie exchanged relieved looks, and then he spoke for both of them. "Six months? That's not bad."

"Could be less, or more. I'm totally guessing based on what I think. I have to warn you, though, that sometimes what I think is a considerable distance from accurate."

Ray stood, breakfast bill in hand. "One last thing," he added. "Can you keep our names out of it for now? I work for the power company and Melanie has her own business. You know how people talk. If it gets out and around in a big way that we're involved with UFOs and aliens. . . ."

"I can keep you out of it until we have an answer. If it turns out to actually be an alien, then of course there's no way to avoid the exposure. You'll have to step forward."

"If that happens," Melanie fretted, "our livelihood might be on the line, so then there'd *better* be some money in it."

"I don't see how it could be otherwise."

"Fine," Ray said. "Let us know how it goes at Laughlin."

CHAPTER FIVE

LAUGHLIN

My suspicion is that the universe is not only queerer than
we suppose, but queerer than we *can* suppose.
—*J.B.S. Haldane*

The International UFO Congress was the brainchild of
Bob and Teri Brown of Denver, Colorado. They spon-
sored two such gatherings each year—one in summer,
one in winter—held in a large casino in Laughlin, Nevada, a
small town as dedicated to gambling as Las Vegas, but entirely lacking panache. Las Vegas had Wayne Newton and the
Cirque du Soleil; Laughlin had things like Hell's Angels' gatherings and UFO conventions.

Six months earlier, I had attended my first UFO Congress.
It was a full week of activities, going from 9:00 am to 9:00 pm
each day. More than thirty speakers (I was one) gave presentations; "experiencers" gave very personal, often emotional
testimony of encounters or abductions; and films were shown
to compete for an EBE Award. (Extraterrestrial Biological Entity, EBE, is the government's in-house term for aliens. They
use it in classified documents when they write about the enti-
ties they insist don't exist.) For UFO buffs, the EBEs awarded
at Laughlin were the equivalent of a cosmic—and sometimes

comic—Oscar, so they were valued by producers as such.

In the vendor area there were book and video sellers, along with purveyors of a dazzling array of UFO paraphernalia, including my favorite, a woman who sold metal pyramids to wear like a hat so they could pull "pyramid power" directly from the ether into your brain. I never saw her with hers off.

Because half of what I lectured about revolved around hominoid research, I entered that world as a different kind of alien. UFO people tend to be highly skeptical about hominoids, while hominoid people are equally skeptical of UFOs. Both groups are, for the most part, down-to-earth types—except for the singular experience that drives them into one of those two areas of interest. Nearly every UFO buff has actually seen one, or in some way *knows* they are real. Likewise, most hominoid supporters have seen one, or they know someone they trust implicitly who has seen one.

In both groups, relatively few have college degrees and they tend to be politically conservative. That inclines the vast majority of each group to believe whatever they're told by mainstream media, *except* in the area where they *know* the media have it wrong. So because the media tell UFO people hominoids are a fantasy, they usually believe it. And, likewise, because hominoid people are assured UFOs are a similar fantasy, they believe that with equal consistency.

It's schizoid living with one foot planted in both worlds.

By the end of my first Congress, I knew the routine and most of the major players. Now, six months later, I'd return as "one of the gang," with friends made from last time, and also ready to make new acquaintances among a select few hundred who managed, by one means or another, to attend what was more like a twice-yearly family gathering. Many who came had spouses or partners strongly opposed to the other's interest in UFOlogy. This was a way to enjoy a week of mental and emotional freedom—to break away from the routines of life to explore, to learn, and to occasionally meet someone you preferred more than the partner you had.

Put five hundred like-minded people together in one place for a jam-packed week and sparks did tend to fly.

Early Friday evening, February 19, 1999, I checked into the casino's hotel. Tomorrow's kickoff presentation would be at 9:00 am, and that was going to be me. I knew Bob and Teri would be anxious to see me, to make sure I was there on time, feeling okay, and ready to give them a good start to establish a positive tone for the rest of the week.

I went straight to my room to tuck away the skull box, then I went down to the ground floor "convention" area to find my hosts, expecting they would be up to their necks with setting the stage for the presentations. Sure enough, both were directing the efforts of a couple dozen staff converting the casino's gargantuan conference room into the home-for-a-week of this year's winter Congress.

Bob is a calm, soft-spoken, serene-in-a-crisis man, befitting a commercial airline pilot. Teri is a bundle of energy, talent, wit, and charm, able to sweet-talk or crack the whip as occasions required. We'd met at the Ozark conference where I met Dan Alegro, so, as with Dan, we were still in the stage of getting to know each other. But I liked both from the start.

Teri and Bob Brown, founders of the UFO Congress held in Laughlin, Nevada.

"Lloyd!" Bob called when he recognized me walking his way across the yawning space that by tomorrow morning would be filled with hundreds of the chairs workers were unfolding and setting in place as fast as they could get it done. "Good to see you!" he said as we shook hands.

Teri walked over to join us, her smile revealing none of the strain she was under to get everything squared away on time. "You're looking well. Did you have a nice drive out?"

"The operative word would be 'interesting,' nice and interesting. Listen . . . can you two break away for a few minutes? I need to talk with you about something important."

They could see I was serious, so we moved over to some stacked boxes that were a part of the stage and the stage props, lighting system, and sound system being set up under their practiced guidance. I gave them a concise recounting of how I had acquired the skulls crashing the Congress party.

"I want to see them," Bob said without hesitation.

"Me, too," Teri added. "Where are they?"

"In my room, but I don't want to start bandying them around until we have a better idea of what the weird one might be. For now, I think it's best to have a small control group evaluate them in secret."

"What kind of group?" she pressed.

"People who can reliably, intelligently contribute to the discussion. You know me, I'm not really a UFO person. It's never been in my field of interest or study. In that regard, I'm starting out well behind the curve."

"You want us to pick your group?" Bob said, frowning with concern. "I'm not sure we should do that."

"Bob's right," Teri chimed in. "Say we don't choose a certain person, then later they find out we chose someone they think was less deserving than they feel. We made an enemy we don't need. Give us another option."

Now it was my turn to frown. I hadn't considered that even here, in a place like this, politics had to be played. Wherever you go, egos are fragile.

"Here's an idea," Bob said. "We could start with the Congress' Board of Directors—a dozen people. No one can complain if we start there. Then, if they have ideas about who should also be evaluating the skulls, we can invite those people for later viewings. That would probably work."

Teri nodded, so I agreed. "Okay, the Board it is. When?"

Teri ran through the schedule in her head. "Tomorrow, the first day, is always nonstop, but we'll end a bit early on Sunday. How about 9:30 in our suite? Bob and I will tell each Board member about it, then we'll meet in our suite and keep everyone else out for a while. Will that work for you?"

"Sounds fine. Let's do it."

Throughout Saturday I did my best to forget I had the skulls. It was Game Day for me, so I had to totally focus on my presentation and the fallout that always followed. I sold and signed nearly 200 books that day. It was a terrific start to the week, and Sunday seemed just as promising.

Bob and Teri's suite was on the casino's top floor, commanding a pristine view of the Colorado River below and the arid mountains of the Mojave Desert beyond. Laughlin, like Las Vegas, was stranded in the Mojave, but the wide, winding Colorado was the difference. Las Vegas had to import all of its water, while Laughlin had an apparently endless supply. That big, beautiful river fits in that harsh desert like the proverbial diamond in a goat's rump.

The suite was always outfitted to meet the needs of an ongoing klatch of visitors at all hours of the night and day. It was where friends and staff could hang out during down time, where stimulating conversation could usually be found. Drinks, and chips and dips, and plastic cutlery, too. It was rumored that Bob and Teri didn't sleep for the entire week.

Early arrivals to the 9:30 meeting stood around chatting uneasily, wondering what the fuss was about but not trying to circumvent my wish to conceal the skulls until everyone was there. When the last straggler finally arrived, Teri hung a *Do Not Disturb* sign on the suite door and locked it.

Bob placed a card table to one side, then arrayed everyone in a semicircle around it. I stood opposite, facing a group whose love, whose passion, whose hobby was UFOs, aliens, and UFOlogy. They knew the history, details, and pros and cons of all major encounters and incidents. They were *educated* about them, so the air was thick with anticipation.

"Thank you all for coming," I began. "I realize you could be doing other things right now, mostly fun things, but I believe this could be well worth your time and trouble."

I lifted the skulls from their box and explained their background as both made their way from hand to hand. The suite floor was padded with thick carpet, but, all the same, I asked everyone to be careful handling them. Carpet or not, I didn't want to begin my stewardship with a drop.

Right out of the chute, a few were convinced it was a Grey. Others felt it had to be a hybrid between a Grey and a human. That was the first time I heard *hybrid* used in that context, which made more sense than it being solely a Grey. It had the eight skull bones of all humans—a frontal, two sphenoids, two temporals, two parietals, and an occipital—yet they didn't resemble the shape and texture of human counterparts. Each was so wildly skewed from normal, and their weight so vastly reduced, it seemed to be a human-type skull transformed into—or by—something else.

But what?

After an hour of questioning, speculating, and jawboning, the initial group was not much closer to a consensus than when we started. There was agreement on only one thing: Both skulls needed testing by specialists right away, as soon as possible, so we could have results before the end of the week, when the conference would end and everyone would return to their homes.

"Mark Bean may be able to help," someone suggested.

I knew Mark from the last conference. He was a computer graphics designer and website manager whose interest was what he called "electrogravitics." He studied and duplicated the machines and experiments of Nikola Tesla, the great Serbian genius who, in his prime, rivaled Thomas Edison as an inventor. Mark carried an array of Tesla's coiled devices from conference to conference, dazzling audiences by sending bolts of static electricity fifty feet out over their heads. Mark Bean was a consummate showman.

"What can Mark do for us?" I asked.

They told me that he lived in Las Vegas and managed the

websites of two or three hospitals there. They felt he might know specialists who could give expert opinions about the weird skull. With any kind of personal connection to whomever he questioned, he should receive straight answers rather than the mainstream garbage dumped on Ray and Melanie.

That made sense. "Okay, I'll talk to him tomorrow."

"Don't forget," Bob cautioned, "he's scheduled to perform at the banquet Saturday night. He can't miss that."

The next morning I found Mark Bean and his partner, Kris Phelps, just finishing breakfast in the casino's café.

Mark was a wry, twinkle-eyed guy with a high forehead that seemed to hold a brain twice the size of everyone else's. His most distinctive feature was a long mane of straight brown hair flowing down to his shoulders and fine enough to stand straight on end when he charged it with his Tesla coils. That was always a highlight of his shows, when he grabbed a charged ball and his hair lifted up and out into a two-foot-long, spark-spewing corona.

Kris, a waitress at Hooters, was by many accounts one of the best looking women to regularly appear on the alternative knowledge circuit. In her late 20s, she was trying to find a partner to settle down with. Mark appeared to fit that bill, so they'd been together for a year and seemed a happy couple.

Mark Bean and Kris Phelps

I went to their table and asked if I could join them. Mark turned to Kris and drolly said, "Well, *this* gets our day off to a rousing start, doesn't it, Hon? Lloyd's here to ask a favor. It reeks off him. Can you smell it?"

Kris was a capable foil for Mark's perpetual antics, so she playfully sniffed the air. "Hmmm . . . bacon?"

"Yes! Right, a favor with bacon. Cheese is extra."

Pulling out a chair to sit down, I asked, "Who told you?"

"Bob called at seven to make sure if I do anything for you, I'm back Saturday by noon. They never sleep, you know."

I nodded. "I've heard that about them."

"Bob said you have a weird skull. Tell us about it."

I related what had happened to me thus far, and then I told him what I hoped to accomplish with his participation.

He winced as he and Kris exchanged a few glances, communicating like Ray and Melanie, in the language of gesture that couples tend to develop. Finally, he said, "You know what you're asking us to do, don't you?"

"Yeah, go back home to Vegas for a few days."

"And miss all this! A week of free fun in a casino beside a beautiful river in the middle of a desert. Vegas is the desert without the river, which has *fish* in it, for God's sake! We've looked forward to this for six months."

Kris nodded emphatically. "Fish, Lloyd, *big* ones."

"You want big? This skull could be history as big as history can be made. I think that's worth whatever anyone can do to contribute to figuring out what it is."

Mark leaned across the table to show me he was taking my grumble seriously. "All kidding aside, Lloyd, what do you think it really is—honestly?"

I shrugged. "Probably 20 percent it's alien, 80 percent a deformity of some kind. But that's mostly because I know so little about it at this point."

"You're in a casino! Does 'stacked deck' sound familiar?"

"We need to be *sure*, Mark. We need to be positive."

He turned toward Kris, whose brow was creased with concern. "What do you say, Hon? Want to help Lloyd out?"

She glared at me. "Today has two speakers I want to see, Jim Marrs and Alan Alford. Alan is at four. If we leave after him, we're home tonight. And I want to be back on Friday morning, no matter what. Will three days be enough?"

That seemed enough to me. "Whatever you can spare."

Three days passed with steady seepage of word that I was in possession of a weird alien skull. I was continually button-holed by people who had heard about it, one way or another, from the dozen Directors at the original meeting, a group that clearly didn't take to heart my request to keep a tight lid on who learned about the skull. Most people asked to see it, even though word was out that Mark Bean had taken it to Las Vegas for a series of evaluations by specialists. They also knew it would be back in Laughlin before Mark's presentation at the closing banquet on Saturday.

"Can we see it then?" they would invariably ask.

"That depends," I would invariably answer.

The truth was, I couldn't decide what to do about displaying it. I had already promised Ray and Melanie I'd keep it private until we had at least a marginal handle on what it was, or might be. But as word of the skulls filtered out to the Congress attendees, advice began to filter back to me, sometimes elliptically, but most times straight to my face.

"Watch out for Men In Black. I'm serious, man, they'll sneak up on you and make you disappear. Gone. *Poof!*"

"The government won't let you keep it. It's too incriminating for them. Heads would have to roll."

"At every conference like this, the government sends agents to infiltrate us. You know who they are, don't you? Everyone does, just ask around."

"Majestic 12 will steal it from you. That's their job."

"You're gonna end up dead over this, Lloyd. Give it back while you still can. While you have it, your life is in danger."

I'm not exaggerating here—I'm being conservative. And the truth is, they were truly scared, rightly or wrongly, and they scared the hell out of me. They really did. I was an utter neophyte regarding all matters related to UFOs, aliens, Men In Black, Majestic 12, and the various other "black" or "deep cover" designations for groups that allegedly did the dirty work necessary to keep UFOs and aliens from being taken seriously by anyone anywhere. These nefarious efforts, my confidants insisted, were spearheaded by the world's most powerful governments, foremost of which was our own.

By Friday, when Mark and Kris returned, skulls in tow, my pucker factor was well into the red zone. Before I even heard their report, I knew that if I announced to the crowd what I had, and what I wanted to do with it, my fate could be sealed. However, if I held back and didn't formally announce it, I could try to find out more about it before I stuck my neck out far enough to get my head chopped off.

On the other hand, if anybody wanted my head—or the weird skull—on a platter, it would be best if many people knew about it. A mere dozen or two could be shut up by threats, intimidation, or outright elimination. Five hundred—that would be a logistical nightmare!

"Irish foreplay," Mark said, as he and Kris entered my room on the casino's fifth floor. Last evening they had left a message saying they'd meet me there at 11:45 am. It was 11:55 when they knocked. "Know what that is?"

"Irish foreplay?" *What does that have to do with—?*

Kris spoke in a heavy Irish brogue. "Brace ye'sef, Rosie!"

"For good news?" I asked hesitantly. "Or bad?"

"Both," Mark replied, handing over the skull box. "A lot happened and nobody's sure of anything, but at least we have some solid data to consider."

Kris opened her purse to remove a small envelope. I took it and found a thin stack of black-and-white photos attached at the edges and folded back-to-front like an accordion. The top one was an X-ray of some kind.

"Video fluoroscopy," Mark explained.

"Show-off," Kris muttered. "They're X-rays."

"Fancy X-rays," Mark countered. "Very upscale."

I unfolded the string and could see it was a shot-by-shot comparison between the adult skull and the weird one. I had no idea what I was supposed to be seeing.

Mark and I took the two seats at the small table in my room. Kris sat in the TV chair in the corner. I spread all the X-rays out on the table. "Tell me what happened," I said.

"First day, not much. Couldn't find anybody interested. As soon as I said, 'weird skull,' or 'could be alien,' they all had something else to do. The next day, Wednesday, I got smart. I started saying I had an old human skull I needed an opinion about. Arranged two appointments after that."

"Almost three," Kris noted. "One fellow had to cancel."

Mark shrugged. "I think that was legitimate. Anyway, at first both guys were like, 'Wow! What *is* this, man? I've never seen anything like it!' But when I told them the story you told us, a chill set in and they started hemming and hawing. It was strange because I know both guys, they're well-respected eye, ear, nose, and throat surgeons."

"They should have known what they were looking at."

"At first they did—you could see it in how they behaved."

He glanced at Kris, who nodded. "They were jazzed."

"When the chill set in, they went strictly by the book, insisting it had to be a deformity caused by cradleboarding and hydrocephaly. When I told them the reasons you don't think it's either one of those, they came back at me with other disorders like Trisomy 13 and Apert's Syndrome."

"Don't forget Crouzon's Syndrome, too," Kris added.

I'd never heard of any of them. "What are those?"

"They didn't explain them, they just said it was probably caused by one of them. Their main message was: Don't worry, it's a typical deformity, go home and forget about it."

I tapped the X-rays. "So where do these fit in?"

"Yesterday," Mark explained.

"Pay dirt," Kris chimed in.

"A friend, Greg, is an X-ray tech at a big hospital. He's a good guy, not a doctor, so I figured he might give us some straight answers. Sure enough, he's into UFOs. He told us he saw one when he was a kid, so he was okay with the idea that the skull might be from an alien or even a hybrid."

"He didn't know about the Congress being here," Kris added. "Now he's interested in coming to it sometime."

"That's irrelevant to what he did for us," Mark resumed. "He ran both skulls through his machines, like he does all the time, and these are what came out. Blew his mind, so he went out and asked four staff doctors to give opinions. Didn't show them the weird skull, just its X-rays. Told 'em it was like a quiz he was working on, like a test."

"He suckered them," Kris added.

"What'd they say?"

"The whole back-of-the-head thing confused them. They were sure the only way to get so much of it so flat was some kind of head binding, but it can't be regular cradleboarding because it would tear the neck muscles loose to go down so low. They said something else had to shape it like that."

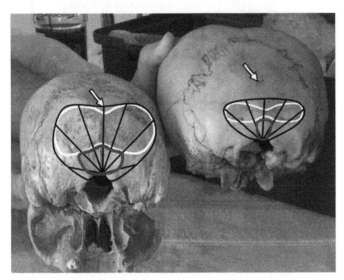

Flat rear in SWFS (left) stops where its neck muscles attach. The SC (right) has smaller attachments, and flattening is comprehensive. Arrows point to same areas on occipitals.

"All right, good to know. Was anything else significant?"

"Probably the main thing they noticed was no trace of any frontal sinuses. In a normal human skull, here, the cauli-flower looking thing between the eye sockets is what sinuses look like in typical X-rays. Now, look at this one."

The weird skull had not a trace of sinus, not even vestigial buds. This was a being born to live without sinuses, which I'd never imagined because I felt they were essential.

"The docs said it's entirely possible to live with extra small, or even dysfunctional sinuses, but to be *missing*. . . . That's an extremely rare disorder, maybe one in a million."

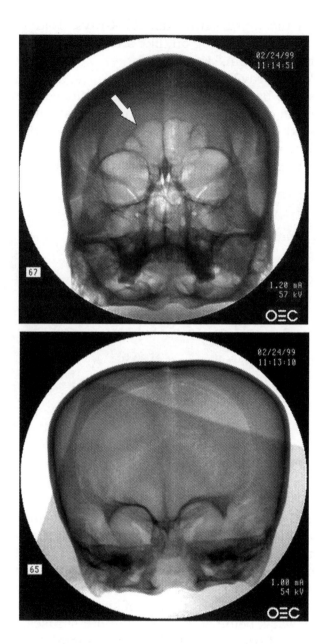

In topmost photo, a fluoroscopic X-ray shows normal sinuses between the eye sockets of SFWS. Directly below, none are present in SC. [Rectangular shadow in SC's X-ray is an artifact of the fluoroscopy process, and does not reflect a part of the skull.]

I pointed to the area of the neck openings, where the spines entered the skulls. "What about this?"

"The neck holes—foramen magnum. Right, Kris?"

"You got it, Babe."

"The weird one has a piece missing, the front part, but the hole is what counts. It's shifted forward a full inch, way past what's acceptable for normal variation. What that did, though, was to give the skull a new way to balance itself directly over its center of gravity—front and back, left and right—like a lumpy golf ball balanced on a tee."

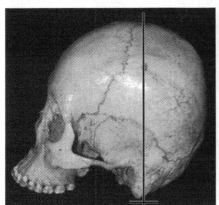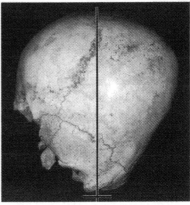

Balance points on both skulls. The SFWS (left) has a normal balance point noticeably rear of center. At right, SC's balance point is well centered.

"Mark hates golf," Kris noted.

He ignored her sidebar and plowed on. "The weird one also had a really thin neck. Check the adult one. See? Its neck connects here at that bump on the back of the skull and sweeps around to the bottom of the ears. It covers a big area on the bottom of that skull. But the weird one, its bump is gone. Missing! Hadn't you noticed that?"

"The inion, right. I noticed it didn't have one, but I didn't know what that meant."

"Neither did they, but it's a big deal. I mean, shifting the entire neck connection down and forward, cutting the size of the neck in half, what's up with that? That's not normal! *Nothing* about this skull is normal!"

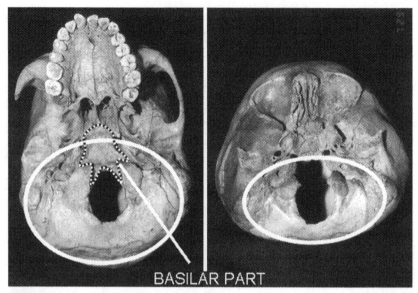

BASILAR PART

Underside view shows the size of a normal human neck (left), and the reduced size and altered shape of the SC neck (right). Note SC is missing its basilar part, which in normal humans is not permanently attached until age 20 to 25.

"He gets carried away," Kris muttered.

"As long as it's not feet first!" he quipped.

"Everyone describes Grey necks as thin like that," I said.

"It gets better," Mark resumed, pointing to the shot of both skulls in profile. "See how much brighter the weird one is? That means its bone is thinner because more light is passing through it. Look at it! It's thin everywhere, uniformly, all over! That's off the scale of variation, even for a young kid."

Ray and Melanie believed it was a child because the tips of new teeth were visible in the three empty sockets in front of the two teeth anchored in the maxilla piece. "The front holes look like they have new teeth waiting, don't they?"

Mark nodded and pointed to the maxilla X-ray. "Greg had no doubt these are new teeth embedded in the upper jaw, ready to come down. So even though the visible teeth look pretty worn down from what baby teeth should look like, he says they have to be from a kid around five, maybe six."

"Whewwww!" I exhaled. This was all coming at me like a water cannon. "You two sure didn't waste your time!"

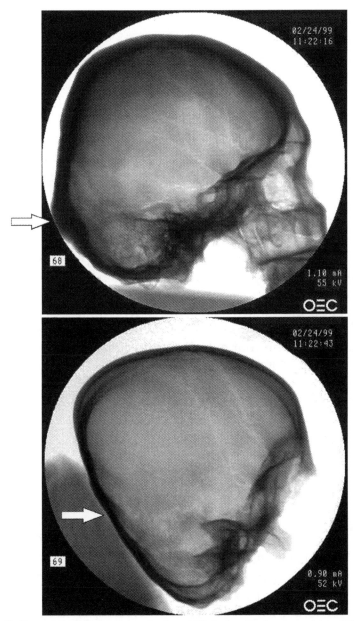

Profile X-rays of SFWS (topmost view) and SC (below it). Arrows point to SFWS's inion, and to area where SC's inion is missing. SC's lighter color shows uniformly thin bone in all quadrants. SFWS's bone is darker and thus thicker. Visible tracks are of veins on the surfaces of their brains that etched patterns onto the inner surfaces of their craniums.

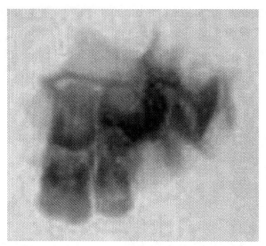

X-ray of SC maxilla showing unerupted teeth. [This was taken using the same equipment used to make the full skull X-rays, and it is shown in its original low clarity.]

"One last thing," Mark said, "and maybe the best part. The brain is extra big. With no sinuses, no deep eye sockets, and no pressure causing that extra bulge in the upper rear. . . ."

"The parietals," Kris put in.

"Right, the parietals. With all that bulging back there, it just has to have a heck of a bigger brain."

Hmmmm. On Saturday, I'd purchased a copy of Colonel Philip Corso's *The Day After Roswell,* and I'd been reading it in every spare moment, starting the long process of getting myself up to speed about UFOs. I was well into marking it with a highlighter, as I did most of what I read,

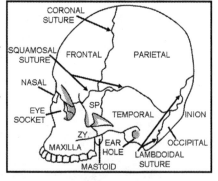

and I recalled a passage at the end of Chapter Six. It said medical examiners working on the alien bodies from the Roswell crash had discovered that EBE brains were significantly larger than normal human brains—*but not at all unlike ours.*

"Did Greg's friends mention hydrocephaly?" I asked.

"No, because hydrocephaly doesn't remove sinuses, it doesn't reduce a neck and shift its placement, it doesn't thin bone uniformly, and it doesn't crease the parietals, does it?"

He had me there. "Good points."

"Absolutely," Mark said with a curt nod. "So the bottom line for Greg is that he thinks it's every bit as strange as you thought it might be. He said if it were up to him, he'd mount a serious, all-out investigation. It's *that* weird."

"You think he'd be willing to help me do that?"

Mark's mane of hair rippled as he shook his head vigorously. "Oh, no, man, don't even think that! He was clear that we can use the X-rays, but we can't use his name as a reference or he'd risk losing his job for helping us."

Frowning, Kris added, "Don't expect much help from doctors, Lloyd. This kind of stuff seems very scary for them."

"It's a real hot potato," Mark concluded.

I was already feeling the heat. . . .

Tapped phones be damned, I called Ray and Melanie to discuss the situation. I told them to get on separate lines so we could have a three-way conversation, then I explained Mark's results in Las Vegas. I said much of the Congress was now waiting for me to introduce the weird skull so everyone could see it before the Saturday night banquet's wrap-up.

"I know we agreed to avoid showing it publicly before we had a solid notion of what it is," I concluded. "But on the other hand, the cat is definitely out of the bag."

"Any ideas about how to present it properly?" Ray asked.

"He means, in its best light," gracious Melanie chimed in. "So you don't come off like just another nut case."

That was a real problem in our field, the fact that most of what circulated in it was, at bottom, humbug. Picking wheat from chaff was a constant challenge for everyone.

"I can't do anything fancy because all we have to illustrate it are the X-rays. Those should be enough, though. I have to tell you, they're pretty doggone impressive."

"What do you think might happen when people know for sure that you have the skulls with you?" Ray wondered.

I laughed. "It won't be making the cover of *Time!*"

"That's true," Melanie agreed, "but you'll also be vulnerable from then on. Remember, the government denies UFOs exist. As long as anything new about it sounds kooky, they let it slide. But anything serious, like this, they'll drop on it like a two-ton tick. You *must* be careful. You could lose it."

I was already deeply concerned about the same thing, but I'd worked myself between the well-known rock and the hard place. A hundred or more attendees now knew I had them, and among those, I assumed, would be any government agents sent to infiltrate and monitor the Congress' activities. I was becoming more and more swayed by grim suggestions that the government was ruthless in how it handled all things related to serious UFO evidence.

No Laughlin confidants could direct me to anyone with first-hand experience regarding the mysterious Men In Black (who, according to some in the UFO field, might also be aliens rather than government agents), or any of the other groups I was being told would soon have me on their "most wanted" list. However, all those I talked to assured me they knew at least one person who had such an experience, or knew someone who knew someone who had. In no case was there any doubt that if you had an object or artifact that even *might* be from a UFO, it *would* be confiscated by the government.

So where did a weird skull fit into this? I didn't know, but I sure didn't want to find out the hard way, which was what I kept hearing in the three days Mark and Kris were gone to Las Vegas. Now I had to face a grim possibility, if not a probability. If the skulls seemed to threaten the government's "official" denial of UFOs and/or alien reality, I could expect them to be confiscated soon. So how was I supposed to deal with that? What could I do to prevent it? These questions drummed through my head as I wrestled with Mark's news.

The rock was definitely squeezing the hard place.

"Bob, Teri . . . I'm afraid I have to ask you for some time tomorrow to introduce the skulls to the people here."

It was late Friday night and I had arranged to meet them in their suite. As usual, staff were in the living room, so we met alone in their bedroom. The bed was made and the room orderly, reinforcing the rumor that they never slept during this week. I had always dismissed it, but maybe it *was* true.

Teri shook her head. "The speaking schedule is locked in, you know that. We can't ask anyone to give up stage time."

I knew this was a problem, so I had a tentative solution. "How about during the banquet, in the after-dinner slot?"

Bob blanched. "No, no no! Too many people want to hear another one of your stories. Let's think of something else."

"Listen, Bob," I said evenly, "I'm not in any mood to be funny. This is extremely serious business for our field and you know it. Let's use that time wisely."

Their disappointment was genuine, and so was mine.

At the end of the previous Congress, my first, I was asked to give a twenty-minute after-dinner address at the closing banquet. Bob and Teri requested something lighthearted, with a smidge of humor if possible, as a counterpoint to the overall solemnity of the week's activities. I decided to tell the story of something that had happened to me earlier that year, something the crowd could relate to.

Art Bell was the top radio personality in our field with his *Coast-to-Coast AM* late-night radio talk show. He was also a huge fan of the Green Bay Packer quarterback, Brett Favre, at that time the most valuable player in the NFL. Art had purchased a genuine $200 Packer helmet and wanted it signed by Favre, so he sent it to the head of his fan club in New Orleans, a friend of mine named William Max.

Art's assumption was that because Favre came from a small Mississippi town not far from New Orleans, it would be no trouble for William to drive over and get it signed. William didn't know where to begin solving this problem, so he called me for advice because he knew I had played football in college at Tulane University in New Orleans.

I had no more of an idea of how to contact Favre than William did, so we set about tracking him down. We found he'd soon be at a golf tournament in Biloxi, so we decided to drive over and ask for the favor. And then the fun began.

What followed was a hilarious series of off-the-cuff improvisations we strung together up to the moment Favre signed the helmet. (An 18-minute video of me telling the story is in my personal website's *Humor* section at *www.lloydpye.com.*)

When telling that story, I had the great advantage of being from the south, where colorful yarn spinning is an art form. I was no artist at it; I knew a dozen people who did it better than I ever could. But usually I'm good enough, and that night I happened to be a little better than good enough.

The success of that story secured my status as an after-dinner raconteur, so I'd been asked to do it again this time. In fact, during the week more people asked me about what kind of story I'd tell at the banquet than asked me about the skull in my possession. By any measure, this was a difficult call; however, I really was in no mood to be funny.

"People will understand," I assured Bob and Teri. "When they grasp the importance of the skull and see the X-rays, they'll be okay with missing out on a few laughs."

Teri still shook her head. "We sold several tickets on the strength of our promise that you'd tell another funny story at the banquet. I don't think we can renege on that."

"All they'll want to grasp is our necks," Bob said sourly.

"We *have* to do it this way," I insisted. "It's not a time for being funny. The rubber is meeting the road."

They knew I was right, so in the end we agreed that I should use my speaking slot to introduce the Congress to the weird skull. With only the two skulls and the X-rays for support, twenty minutes would be enough.

"What will you call it?" Teri asked.

I shrugged. "A weird skull. That's what it is."

Bob's promoter hackles promptly rose. "You can't just call it a weird skull, Lloyd. You need to hype it a bit or no one will pay any attention to it."

"You need something with spark," Teri added, "*pizzazz.*"

"They *will* pay attention," I insisted. "How can they not?"

Teri patted my arm. "Give it some thought, Hon. It needs a memorable name, something people can relate to."

I lapsed into silence, thinking, until I recalled something from my conversation with Mark and Kris. "One of Mark's sources in Las Vegas said it's the skull of a child, for sure."

"Child of the cosmos?" Teri suggested. "Or how about child from the stars? Something along those lines. . . ."

I shook my head. "Shouldn't we be more conservative until we have more information? I mean, the odds are a mile long against it actually being from an alien."

"Not among UFO fans," Bob said. "Here it needs a name that grabs people. Out there in the world, among scientists, you can call it a weird skull. But if you want *our* people to get enthused about it, you should give it a catchy name."

Made sense to me. "Okay, what about 'Starchild'?"

Bob and Teri glanced at each other, eyebrows raised.

"Starchild. . . ." Teri mused. "I like it. That could work."

"Okay," I said. "For now, to introduce it to the crowd here, I'll call it the Starchild skull."

Despite intending to keep that name confined to Congress attendees, once we agreed on it and I spent some time using it, I never thought of the skull as anything else—*Starchild.*

The UFO Congress banquet is semiformal. Men dress in coats and ties, women wear cocktail dresses with full war paint and jewelry. Large round tables seating a dozen people are spread across the presentation room. Tickets are pricey, but most attendees end the week at the banquet because, in a word, it's fun. You choose your own seat with people you know or came to know during the week. Food is good and ample. Jokes and skits are mixed between the EBE awards for excellence in UFO-related films and videos for the year. Filmmakers take the EBEs seriously because no one else takes *them* seriously, so competition is stiff. If you can't have fun at the UFO Congress banquet, you're just not trying.

Into that buoyant atmosphere I came with the skull box. I stowed it under the podium so I could keep a close watch on it until my turn to speak, which ironically came after Mark Bean put on another of his spark-spewing extravaganzas. His act was tough to follow, but I did it last year with my funny Brett Favre story. This time, I had nothing funny to say.

Bob introduced me by explaining to everyone that I had something very special to share with them, something that would prevent me from telling another funny story, but he promised I would do that at the next Congress. By the time I took the podium, the cavernous room was abuzz. Most heads twisted and turned to glance at tablemates with confusion. Everywhere, shrugs and raspy mutterings were stilled by the many who now knew what I planned to do.

Leaving Ray and Melanie out of it, as per their wishes, I began by briefly explaining the background of the skulls as I understood it, and how they came into my control. Then I contrasted the adult with what I now termed the Starchild, explaining my goals for it, what I hoped to do with it and for it, and how I hoped to go about getting that done.

Next, I held both skulls high for everyone to see, while a video recorder blew them up to a huge size on the screen behind me. Nobody could fail to see what the Starchild was and what it looked like being recorded for posterity.

A new chapter of its provenance had begun.

When everyone had a clear image of the skulls, we put the X-rays on the screen and I discussed each pair of shots, repeating what Mark and Kris had told me their experts said. I gave each as thorough an explanation as was possible in the time allotted. As I neared my conclusion, I had something left to say that I felt certain would not be well received. However, it was painfully stuck in my craw, and I wasn't leaving Laughlin without trying to dislodge it.

"In closing, let me assure you I'm fully aware of what an important responsibility this is. I accept it with as much appreciation and seriousness as any of you would if you were standing in my shoes. However, *I'm* standing in my shoes, so I have to tell you something, right up front, that all of you might not approve of. I am *not* prepared to die for this. I'm willing to work hard for it, to work diligently, but not to die.

"I say that because during this week I've talked to several of you who have assured me that the government will be bound and determined to relieve me of these skulls as soon as they find out I have them. Those who told me that know who they are, and many others would have told me the same thing if, before this evening, you knew I had them. Am I right?"

Heads bobbed at tables all around the room.

"Everyone I talked to seems to agree that this course of action I'm setting off on is foolish at best, and it could easily become downright dangerous. That being the case, I want to announce something here and now to all of you."

I paused to look at two men I had been assured were government agents who monitored activities in this field. Between them, they attended all major conventions around the world. I'd seen them a few times during my travels, and I liked both. They were extremely personable, and able to drink anyone under the table. Having been an intelligence agent, I knew what that was about. We were trained to use alcohol to obtain information we wanted because, invariably, it worked.

"I won't fight to hold onto these skulls. If the government wants them, they can have them. They don't have to kill me, or beat me up. If they ask me for them, I'll hand them over."

I said that looking first at one man, then at the other, then back to the wide-eyed audience sitting with mouths agape.

"There is *no* need for them to kill me. I won't be armed, I don't carry weapons. If I'm found dead with a gun on me or nearby, it was planted. I won't fight for this, so if anything bad happens to me, understand that it didn't *have* to happen. It happened because they want to teach *you* a lesson—each of you. Not me, I'll be dead, but they'll be trying to intimidate all of *you*. If intimidation is their game, don't let them win."

Heads shook or nodded slowly as I wrapped it up.

"In six months many of you will meet here again. I hope to join you. But if I don't make it back, I have a favor to ask. Make sure the film of this, of what I just told you, finds its way to somebody at *60 Minutes*. Okay?"

Somber nodding throughout the room.

"Thank you for your time, your attention, and your support. Without it, I couldn't do what I have to do with this."

I've been applauded in my life, but never like that.

CHAPTER SIX

VEGAS, SEDONA, DENVER

The great obstacle to discovery is not ignorance—it is the illusion of knowledge.

—Daniel J. Boorstein

W hen the UFO Congress ended, I still had to give four lectures in four states before returning home. The next would be in Sedona, Arizona, the following weekend, so I had a full week to cool my heels. To fill downtime I'd usually visit with friends in the area where I happened to be, staying a night or two with each one—just enough to keep from becoming a *real* pest. Everyone I barged in on seemed to realize I was on a mission of sorts, so it was easier for them to humor me than to try to talk me out of it.

For this particular week I had arranged an important—and very much desired—destination: Mark Bean and Kris Phelps had invited me to stay with them in Las Vegas. Before we left Laughlin, Mark had agreed to be my partner in the Starchild effort. He kept harping: "You cost us half of a paid-for, one-week vacation—you owe us!" It was said in fun, though, because he and Kris were now hooked on this strange, unfolding mystery. We all wanted to *know.*

Their Las Vegas home was in a country club built around a golf course. On the surface, it was an odd choice of domicile for someone with antipathy toward golf, but, as usual with Mark, there was method behind the seeming madness. The country club was well guarded with a permanent entrance and exit gate, which was important because of the almost inestimable value of his Tesla machines, all of which he had more or less built by hand to Tesla's exacting specifications.

Their home adjoined the sixth fairway. As soon as my belongings were tucked into a spare bedroom, I started chiding them about the ostentatious luxury of their surroundings.

"Listen, man, it's a *war zone* in here," Mark insisted.

"We have to wear safety helmets in the kitchen," Kris chimed in, pointing to a plastic hardhat on a nearby hook.

"Twice they've launched grenades through the window!" Mark clapped his hands loudly. "*Ka-boom! Ka-boom!*"

"*Crash!*" Kris added. "Broken glass everywhere! One grenade went right into the aquarium, honest!"

"Was it a hole-in-one?"

"I doubt it was even a mulligan!" Mark chortled. "People around here take golf *seriously!*"

Since Mark didn't golf and I'd never been interested in it, we let it drop and got down to the business at hand. I was there to accomplish two things: (1) Mark had convinced me that the effort with the skull needed a website, and he offered his professional services for free; and (2) we needed more input from specialists regarding the skulls. Doctors' opinions were certainly valuable for assessing deformity, but I felt anthropologists might be less intimidated by, and more curious about, the Starchild's high degree of . . . unusualness.

Naivete is, obviously, a hallmark of the uninformed.

The University of Nevada at Las Vegas (UNLV) is a major university serving a student community of over 25,000. I called its anthropology department and asked to speak with the person most familiar with human skulls. They put me through to a fellow I'll call "Roger White." I told him I had a pair of human skulls I'd like him to examine. He responded by asking several pointed questions about me: Who was I? For whom did I work? How did I acquire the skulls? Why did I need them evaluated? I tried to heed Mark and Kris' warning to avoid using buzzwords like "UFO" or "alien," but it turned out to be more difficult than it sounded.

I had no university affiliation, no medical certificate, and no forensic credentials. I was a guy calling out of nowhere to ask a local anthropologist to examine a pair of skulls. I just kept telling him it was a personal request, a personal issue, and that I didn't know where else to turn. Finally, he gave in and agreed to meet me, probably to find out more about me than the skulls I wanted him to examine. He may have thought I was some kind of ghoul.

Mark went with me. Kris' work at Hooters was tightly scheduled, but Mark's computer expertise allowed him to work at home, so his time was his own. We drove to the UNLV campus and found it huge, but the directions Roger had provided quickly brought us to the building where we arranged to meet. The room was a typical laboratory holding pieces of equipment laid out on several tables. Skulls and other aspects of human and animal anatomy were arrayed on shelves along the walls. This was the real deal, no doubt about it. These people studied bones—big time.

Roger, we found, looked to be in his mid-thirties, but he was lean and wiry and could have been older. He'd brought a backup in the person of "Dr. Green," an older woman with short gray hair and crinkled, wary eyes. Roger was cordial and properly polite, but Dr. Green made it clear from the get-go that she wasn't pleased about us being there.

Without much fanfare, I pulled the two skulls out of the box and put them on one of the tables. Both professors stared, momentarily transfixed.

"That's clearly a cradleboarded adult," Roger said, pointing at the normal skull. "Probably a female."

Dr. Green nodded, adding, "Very likely an adult female, Roger. But the other one is why they're here." She shifted a cold, impatient gaze toward me. "Am I right, Mr. Pye?"

"Exactly, Dr. Green. Can you tell us anything about it?"

She stepped to the table and picked it up, hefting it as anyone would, then she handed it to Roger, who hefted it the same way. They shared a brief glance of confusion.

"Desiccation?" he suggested, to explain its lightness.

Her head shook. "I'm not sure. It's definitely unusual. But I think we can agree that it's probably hydrocephalic. That invariably causes some degree of bone thinning."

"Only in the areas that expand from the outward pressure," Roger noted. "This bone seems to be thin all over."

Dr. Green nodded, then changed the subject. "Notice the lack of a brow ridge, and the narrowing of the eye orbits."

"Their shallowness, too," he added. "Highly unusual."

Dr. Green nodded again, this time a bit grudgingly. "The cradleboarding is as extensive as I've ever seen."

Roger turned to me and said, "Where is it from?"

"Mexico, a hundred miles southwest of Chihuahua."

Dr. Green flashed a triumphant smile. "Well, then, that explains the cradleboarding. It was a common practice of the Tarahumara. They still do it sometimes today."

"Do you have any idea when they died?" Roger asked.

"They were found around 1930, so we know death was sometime earlier. But when, specifically, we don't know."

"Carbon 14 will tell you that," Dr. Green advised.

"We know, but we—Mark and I—are just getting started with trying to figure this out. You're actually our first stop."

"We can be your last, too," Dr. Green said, with a note of snappish dismissal. "It's almost certainly a cradleboarded child suffering from hydrocephaly."

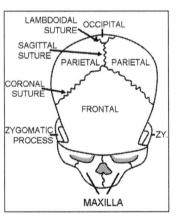

Suddenly, Mark Bean spoke up. "If it were a true hydrocephalic, Dr. Green, would that crease be in there between the parietals?"

I was cue-ball bald, with a salt-and-pepper, close-cropped fringe. Mark's shoulder-length mane was exactly opposite. He was young, I was old. We were such an odd pair, they had been ignoring him. Now they had to regard him as what he was—an intelligent participant to be dealt with.

His pointed question forced Dr. Green to become even more cautious. "If the sagittal suture is fused, such a crease would be expected. I admit, though, this doesn't look fused."

I locked eyes with her. "In this case, a guess isn't good enough. We need an answer we can count on."

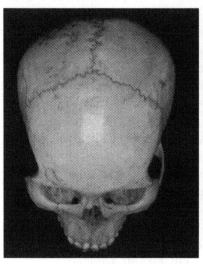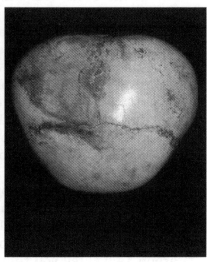

Top views showing "crease" in Starchild's rear crown (right).

Drs. White and Green paused a moment to consider their options, while Mark stole a glance at me, delightedly lifting his eyebrows in a Groucho-like, *Way to go!* leer. I made a hand motion that told him to knock it off. This was serious.

"Listen, Mr. Pye," Dr. Green said firmly, proving she was in charge. "We can't give you the kind of specificity you want. We can only confirm that this skull is very typical of cradle-boarding, as is the other one." She pointed over to it. "Flattening the occipital to *any* degree is a common sign of head binding. There should be no argument about that."

"Only if the bone is reshaped to, say, the smoothness of this tabletop." I rapped on it. "The other skull shows that. Feel it—flat as a board. But the strange one isn't smooth. Flat, yes, extensively, but with its natural convolutions intact. So unless I'm way off base here, it seems safe to assume this skull wasn't flattened by traditional binding."

"It's obvious, Mr. Pye," Dr. Green said testily, "that you know your way around skulls. So why are you here discussing these with us? You seem to have your own answers."

"I don't mean to be argumentative, Dr. Green. What I'm hoping to find are answers that clarify what I know about skulls, and the number of odd things I see in this one."

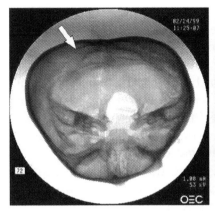 

X-ray of bottom view and rear view of SC showing occipital convolutions.

Still agitated, Dr. Green said, "You shouldn't rule out cradleboarded hydrocephaly because that makes sense in the context of a skull found in 1930 in Mexico. Other options are, of course, available. Various types of genetic deformity may have played a role. These are not our areas of expertise. But whatever deformity it might be, a deformity it surely is."

"I think that's the bottom line here," Roger put in, as if trying to separate us. "It's *some* kind of deformity, whether you accept cradleboarding and hydrocephaly or not."

"Can't you even *imagine* anything else?" Mark put in.

Dr. Green's posture bristled with indignity. "There is nothing else for either of us *to* imagine. It's a deformity—period. Now, if you two will excuse me, I have to go."

That was the end of it, just that quick. We said muted thanks, put the skulls in the box, and were soon out in the blinding glare of mid-afternoon in Las Vegas in early March.

"Holy Maloley!" Mark muttered, using his unique variation on the old *Holy Moly!* I remembered from my youth. "That didn't go well at all, did it?"

"No, it didn't, but we know more than we did going in."

"Yeah? What?"

"We're in over our heads."
"You think so?"
"Try to touch bottom."

Creating a website was our next order of business. Mark went to work shaping it, while I got busy writing the words to fill it. His desk area was jam packed with the latest gadgets and gizmos, like a space capsule. From where he sat he could almost launch himself to the moon. I worked at an old desktop computer, on which I wrote a long essay describing everything we knew about the Starchild skull at that early juncture. I closed it out with a promise to regularly update it as we progressed with our investigation.

Next, as with Bob and Teri, came the problem of naming the website. "What about *Starchild Project*?" Mark suggested.

"As good as anything I can think of. Let's go with it."

So it became, and remains, *www.starchildproject.com.*

By Wednesday morning we were up and running on the Internet, not quite two weeks since Ray and Melanie first showed me the skulls. Things were happening at dizzying speeds, but I couldn't slow down. I'd stepped on a treadmill, for better or worse, and now I had to find a way to keep up with the flow of events as they came barreling toward me.

"While you're here," Mark said, "we need to get you on television. That should be pretty easy to do."

He spoke as if it were a done deal. "Are you serious?"

"Sure! George Knapp should go for this kind of thing."

"Who's George Knapp?" (Pronounced "nap.")

"Oh. . . ." Mark muttered, sweeping both hands through his mane and tucking it behind his ears, something he did when he wanted a brief respite to gather his thoughts. "You know what Area 51 is, don't you?"

"Of course!" Even outside UFO circles Area 51 was known to be an ultra-secret test facility for advanced military aircraft in an isolated area north of Las Vegas.

"George put Area 51 on the map. Without him, it's still a rumor that nobody in government would admit to."

"No kidding?"

"If George Knapp won't jump at a possible alien skull, no-body in America will. He has *cojones* as big as cantaloupes." I sat stunned, trying to grasp the impact such exposure —even local Las Vegas TV—might have. My breath squeezed out in a whistling wheeze. "Whewww! That's *exactly* what we need to get scientists to pay attention to us, to get them talk-ing about it around their water coolers at work. Do you think George Knapp would actually agree to cover it?"

Mark shrugged. "It can't hurt to call him."

George Knapp, it turned out, was an icon in Las Vegas. He was widely known as the best investigative journalist in Ne-vada, not afraid to step on the largest toes or the most inflated egos. His *cojones* were more like watermelon size, which led to his successful challenge of the Federal government's ab-surd claim that there was no such place as Area 51, when ev-ery day planeloads of people who worked there flew back and forth from a Las Vegas airport the way other workers might take a bus, a subway, or drive to and from work.

Before 1989, few people outside of Nevada had ever heard of Area 51 or Groom Lake. Then Knapp produced a series of news reports that established its credentials as the poorest kept secret in Las Vegas. In those reports were sensational yet substantiated allegations that it was the home of spectac-ular technology, possibly alien technology, and in subsequent years every major news outlet had done follow-up investiga-tions. Thousands of stories in print and broadcast media had been generated, including documentaries, movies, magazine articles, websites, books, TV shows, and more doo-dads and other paraphernalia than one place deserved to be recognized for. In many ways it had become a cottage industry in the state of Nevada. Area 51 was good for business.

Apart from that, George Knapp had spent his working life rubbing the wrong way against Nevada politicos and bigwigs of all stripes, and in a very public manner. Somehow he had lived to tell the tale, which made him fascinating to me be-yond my desire to have him put the Starchild on TV. He was undeniable proof that you could make serious waves and survive the experience. I felt I might be on the cusp of making some serious waves, too, so I wanted to have the benefit of his experience if I could possibly draw it out of him.

Knowing such a man existed and thrived made me feel much better—and safer.

He came to Mark's home Thursday afternoon, trailing a cameraman and a soundman. He was as handsome as a movie star, a cross between Robert Shaw and Albert Finney in their primes. He probably could have done anything in entertainment, so I couldn't imagine why he deliberately chose to annoy the rich and powerful. Yet there he was, in Mark's kitchen, chatting with us as his team set up their equipment and we hoped no stray golf grenades crashed through a window to whack any of us. Luckily, none did.

The interview went well. I sat at the table showing George what was unusual about the skull, comparing it with the adult, letting him hold both so he could assess and comment on the obvious weight difference. He said it would be a three-minute piece, but promised to pack those minutes with nothing but the best of what I'd said. The main point I stressed was that we were seeking help from the scientific community to determine what it was.

When we finished, the cameraman and soundman began packing their equipment. As they did, George gazed at me and arrowed to the heart of the matter. "If this is what you think it is, do you understand what that means?"

"Sure . . . aliens are real."

"Do you know what *that* means?"

"Well, that they're real. What do *you* mean?"

"This is a thing certain people in high places in government will not take lightly. I say this from personal experience. Don't expect to get off clean or easy."

Mark butted in. "What do you think they'll do to him?"

"No idea. They've done a few things to me trying to back me off various stories. But you know, they've never actually hurt me. Not so much as a punch in the nose."

"I'm glad to hear it!" I laughed. "Mine's been broken seven times, and I'm sure it wouldn't appreciate an eighth."

George's eyebrows rose. "I noticed it looked a little bent."

"Three times in schoolyard fights growing up, four times playing football. My nose must have bad karma."

George smiled wide, like the movie star he might have been. "Sounds like you have enough testosterone for this line of work. That's a good thing to have. You'll probably need it."

He stood and we shook hands all around. "After we run this tonight," he said, "I plan to send it out as a national feed. I can't see why it wouldn't be picked up by other stations."

"You mean other CBS affiliates?" (He worked for CBS.)

"Right. It's a station-to-station decision, but we can hope that maybe *The Early Show* will be intrigued enough to give you a call. That's your best bet."

I looked at Mark and Mark looked at me, both of us wide eyed, thinking, *Bingo! A grand slam in our second at-bat.*

Our piece was on the local CBS news that evening at 6:00 pm and again at 11:00 pm. Las Vegas was a city of about 500,000, so I calculated those three-minute segments were watched by at least 50,000 people, many of whom would actually work at Area 51. Mark and I waited all evening, well past midnight, hoping for a call. None came.

No call came from *The Early Show*, either, and no other station anywhere picked up the piece to replay it for their viewers. So there it was, an exciting new discovery by George Knapp, the man who introduced the world to Area 51, and yet *nobody* found it compelling enough to replay it.

In the city of gamblers, what were the odds against that?

The following day I had to leave for my next speaking engagement in Sedona, Arizona. The friend I would stay with was what I jokingly called a "metaphysical wacko," closely allied with the highly spiritual "woo-woo" community of aging hippies and flower children in and around Sedona.

When I arrived, she suggested I allow several local shamans and other "sensitives" with metaphysical "powers" to examine the skull to see what they might be able to discern from it. This highlighted that I was not then, nor had I ever been, especially keen on such activities. However, my friend sold me on a logical argument: "You need to know anything you can find out about that skull, right? Well, what if a spiritual medium can tell you something worthwhile? *What if*, Lloyd? Don't reject it until you try it."

She wouldn't take "no" for an answer, so I agreed to meet with them after my usual presentation about human origins and hominoids, which now included a brief postscript about the Starchild and what I hoped to do with it. Afterward, half-a-dozen local psychics of various stripes came to her home, and they had a long session of trying to do what they called "psychometrizing" both skulls.

Each would join me in a bedroom to privately do their "communing" with the "spirits" surrounding the skull. That way none of the others could interfere with, or influence, their individual efforts. Unfortunately, none arrived at conclusions that impressed me, though it was interesting to hear bright people offering imaginative ideas about what might have brought an adult human and such a strange being together in a mine tunnel to die.

This was the first of what eventually became hands-on analysis by more than fifty sensitives reputed to have an ability to tap into dimensions beyond our own. I heard many colorful, descriptive, often highly emotional tales of woe and grief and fear and terror. A few could have been fashioned into decent Hollywood dramas. I was often spellbound by sheer narrative power; some of those people were absolutely brilliant orators. The defining characteristic for each one, though, was an unwavering belief that their multidimensional perceptions, however they were gained, were *correct*—not in every word, necessarily, but in context.

Ironically, within those fifty stories I could easily have heard, word for word, precisely what happened between the Starchild and the person who died with it. Some said they were mother and child; others that they were lovers; others had it as two males; others, two females. Some "viewed" their deaths as quiet and peaceful, others saw unremitting violence. Some felt the Starchild was a young child; others saw it as an adult, or as an old adult, or a superannuated being hundreds of years old. The range was wide and deep.

What I kept seeking but never found was *consistency*. I

never heard the same story twice! Each resulted from opinion and impression, belief and training, knowledge and prejudice, swirled together in the head and heart of whoever told me a tale. Unlike mainstream scientists, who seemed to have lost the ability to be creative in their perceptions, the sensitives and psychics knew no intellectual bounds and had no emotional hobbles, no creative restraints.

They spoke their minds and were usually entertaining, but I never felt confident any were on target with what the Starchild was or how it died. I gave this as fair and honest a chance as I could. It just didn't work out satisfactorily.

My next stop was Denver, Colorado. The people I stayed with had a friend associated with a local medical school, and that friend recommended me to "Dr. Black," one of the most respected brain specialists in the state. I was assured that if Dr. Black agreed to meet me, he could tell me what was worth knowing about the Starchild's brain. At that point any information about the Starchild would be useful, concerning its brain or otherwise, so a meeting was arranged.

Dr. Black turned out to be a small, balding, soft-spoken man my own age who hadn't taken care of his body, or perhaps the world he inhabited had dried him out like a husk. Dark, heavily bagged eyes projected an expression of what seemed like constant woe. He sat in an incongruously large, deeply padded chair behind an equally large, incongruous desk stuffed into a small office. All of it dwarfed his diminutive body, creating a surreal effect, as if only his intellect should be judged and measured. His size was irrelevant.

We began our introductory chat and I soon realized his words were diametric to his stature and surroundings. He spoke with calm empathy and undoubted wisdom. I found myself liking him, respecting him, without yet engaging him about the reason I was there. He was, simply, a nice guy.

When it came time to get down to business, he asked to see the "odd" skull he was supposed to evaluate. I pulled both from their cardboard box and put them on his desk. He studied them for several seconds, then pointed to the adult and lifted the Starchild to examine it more closely.

"That one is a normal adult; this one is something else."

He wasn't just stating the obvious, he was sinking into "game mode." I'd seen the same focus, the same hunkering down to get fully into a task, in every guy I had ever played sports with. Dr. Black was taking this stone-cold seriously, the first time I'd seen an "expert" do that. He studied it for several seconds, flipping it in his hands, front, back, top, bottom—especially bottom. He stared into the foramen magnum hole, then stuck a forefinger inside to swish around on everything he could reach inside: S*wish* . . . pause . . . *swish.*

Soon he stopped and leaned back in his big chair. "All right, I'm interested. Tell me about it."

At that point it didn't take me long to relay what I knew about it in a succinct litany of bullet points, trying not to waste his time, but wanting to be as thorough as possible.

After I finished, he rose and said, "Come with me. I need to take some measurements. Bring both skulls."

We walked down the hall from his office to a laboratory. Inside it, he went to a cabinet and pulled out a large lidded tin. With the tin was a box of small plastic garbage bags, the kind that go in bathroom wastebaskets. He took out two of those and set them and the tin on a long lab table.

"Please hold the adult skull upside down," he said.

He stuffed the plastic bag into its neck hole, then blew into it as if it were a balloon. He then removed the tin's lid to scoop out tiny dark seeds.

"Flax," he explained, filling the plastic bag inside the skull until no more fit. With the other bag in both hands, he sealed its opening around the neck hole. Then he flipped the skull top-side-up, pouring the flax seeds from one bag into the other. He took the full bag to a vertical glass cylinder marked with ascending lines and numbers, and poured the flax in.

"Twelve hundred cubic centimeters," he said, pointing out the line of measurement to me. "Average for a human in the range of five feet tall. You would be closer to fourteen hundred cc's, the overall average. I would be somewhat less."

He referenced his small stature in such a subtle way, I barely noticed as I watched him pour the flax seeds from the cylinder back into the tin, then he stuffed and blew the first plastic bag into the Starchild's cranium. This time he seemed to load several more scoopfuls of seeds into it than before.

He saw it, too, muttering, "Larger . . . substantially so."

When he poured that bag into the cylinder, it was noticeably higher. We leaned in together to see what it read.

"Sixteen hundred cc's," he said hesitantly, as if unable to accept it, or believe it. "Fully a third larger than the other one. That's highly significant because, as you see, the craniums have roughly the same external volume."

The words in Colonel Corso's book echoed again from one year earlier: *"The brain is bigger in the EBE, but not at all unlike ours."* He was starting to sound like a prophet.

Dr. Black turned to rummage in a drawer, couldn't find what he wanted, and opened another drawer. He lifted a metric ruler to hold against the adult skull. Swiftly taking measurements, he shifted to the Starchild. He then put the ruler down, intently considering what he had just learned.

"Each is close to 15 centimeters in height, width, and depth. Both would fit easily in a cube of that volume."

I had already measured them at 6 inches in those three dimensions. "I know."

"So with two skulls near the same size, one holds 400 cc's more brain. As I said a moment ago, that's a full one-third more—well beyond any degree of variation we might expect."

"I told you about its lack of sinuses, and you can see for yourself the shallow eye sockets and the expansion in the parietals. . . ." I left it hanging.

"Yes, I remember what you said. It's just that I don't see *that* much additional space for *that* much additional brain. There must be hidden nooks and crannies that don't show up clearly in the X-rays. It needs more room."

"Are you saying it's like wall-to-wall brain?"

Amused by the metaphor, he smiled. "Yes, that would be one way of putting it."

"What about the overall thinness of the bone? Could that account for so much extra?"

"Maybe, maybe not. Let's go back to my office."

If I thought I saw his game face before, I really saw it now.

He sat in his big chair fondling the Starchild skull, reaching his small fingers as far inside the foramen hole as possible, wincing as the edges bit into the webbing of skin between his fingers. Eventually, he settled into slowly rubbing back and forth across the inside of the lower portion of the occipital bone, the part that was so unusually flattened. Finally, he put it down on his desk and looked across at me.

"This is *extraordinary*. I don't know how else to say it."

Our roles now reversed. I settled back into the padded wing chair opposite his oversized desk. "Tell me about it."

"It's hard to know where to begin. You've already noticed the lack of an external occipital protuberance, correct?"

I nodded. It was the inion, the bump we all have at the lower rear of our heads. It's where two big neck muscles attach, the demarcation between our neck and cranium.

"Its counterpart inside the skull," he went on, "the internal occipital protuberance, is also missing."

He motioned for me to pick up the adult skull. "Put your finger in the adult's foramen opening. Tell me what you feel across the lower occipital."

I did and soon said, "I feel a vertical crest of bone and four flanges of bone flaring out and away from it on both sides, two left and two right."

Again he smiled at my choice of words. "Flanges, yes, let's call them that. And they're quite distinct, are they not? About a centimeter of elevation, yes?"

A centimeter might have been pushing it, but what did I know? "More or less. . . ."

"Exactly. And yet here—" he lifted the Starchild, "—those flanges are nearly gone . . . more like a few *milli*meters high rather than the full centimeter they should be."

He indicated I should take it and make the comparison for myself. I had already discovered this difference between the skulls without realizing its significance. "You think this adds enough room for the extra brain?"

"No, not at all," he said. "Let's just forget about that for the moment. I don't have even a good speculation for that. What I'm talking about now is the cerebellum."

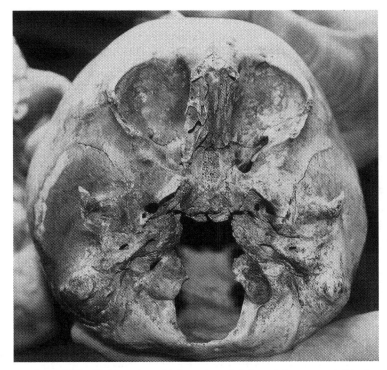

Bottom view of SC with its reduced internal occipital protuberance visible through the foramen magnum (neck hole) and the missing basilar part.

The cerebellum is a fist-sized separate part of the brain resting below and behind the bulk of it—the cerebrum. It's nestled in the curve of the occipital (a curve the Starchild lacks), where it is protected from routine injury by neck muscles attaching to the skull where it joins the cerebrum. Its volume is only 10 percent of the entire brain, yet it contains 80 percent of all the operational neurons. It receives two hundred million input fibers, whereas the incredible optic nerve utilizes only about one million input fibers.

Early research into cerebellar function indicated it was dominant in motor control of the body, directing movements large and small. Modern research has shown it has a broader role in numerous cognitive functions, including attention and the processing of language, music, and other sensory temporal stimuli. In short, it's one of the most important parts of the brain, so anything relating to it in a negative way would heavily impact its owner.

"You see," Dr. Black went on, "I'm having trouble imagining how this individual could have lived long enough to grow primary teeth with its skull and brain so badly misaligned. Let me show you what I mean."

He lifted the adult skull. "In this one, the cerebellum rests here inside the curve of the occipital. Forget that it was cradleboarded; that would have no impact whatever. The part surrounding the cerebellum is never affected by cradleboarding. However, notice the way a typical brain is constructed. The cerebrum, the entire remainder, rests atop it for the lifetime of the individual. It's about three pounds resting on, pressing down on, roughly one-third of a pound. Are you starting to see what I'm getting at?"

"No, not really."

"Those flanges of bone that line the occipital? They're part of a fluid drainage system in the brain, but they also help stabilize the cerebellum in its position so it can keep doing its job over the lifetime of an individual. If not for that solid support, over decades the full weight of the cerebrum could, and very likely would, distort the cerebellum, ultimately causing any number of dysfunctions in cognition and motor skills."

"You'd be screwed, blued, and tattooed."

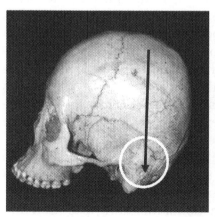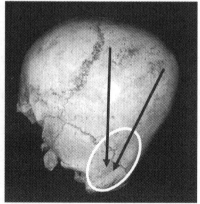

Arrows indicate normal vertical pressure on the SFWS's cerebellum (left), and the greatly increased angular pressure on the SC's more oblong cerebellum (right).

He smiled wryly. "Another astute choice of terms, Mr. Pye, and what I expect was the fate of this skull's owner." He held the Starchild up and made hand gestures to illustrate his points. "In it, the curve of the lower occipital is flattened into this steep angle. In addition, the foramen is shifted forward a bit more than an inch, maybe three centimeters, making that angle steeper still. Now, add to that the additional weight of all the extra brain matter we calculated, resting here in the expanded parietals, pressing down straight toward the foramen opening. See what I mean?"

"The cerebrum was squashing the cerebellum?"

"Had to be!" he exclaimed, becoming a bit agitated. "Its parietals hold far more brain than they should, which we know because we measured it, but all we need to do is look—it's obvious! We can also see the extra volume and weight are canted at a steep angle down toward the foramen, passing *through* the cerebellum—directly through it!"

"Got it," I said, hoping to keep him going.

"Notice where the cerebellum rests. See how pinched it is? Because the occipital is so flat from crown to foramen, there's no curvature for the cerebellum to rest against. It's squeezed into a canted position instead of its usual nearly horizontal position. Even worse, it has a mass of additional weight bearing down on the surface area at its base, which rests flush against the foramen opening. And notice that the surface area bearing all that weight is squeezed into more or less a V shape. . . ." He separated his thumb and forefinger, opening and closing them from a curve to a V.

"I get it," I said, making hand gestures of my own. "It's pinched into an unnatural shape, and those flanges that should give it support aren't there. It's like a gentle slide on the playground, or a scary steep one. Three-year-olds go down the first slide, five-year-olds use the scary one."

Dr. Black blinked in appreciation of that analogy. "Exactly! So what I'm trying to say is that this is a highly improbable brain structure for a human who lived to the age of having primary teeth. I'd expect anyone born with this kind of brain deformity would die shortly after birth."

I didn't know what to say, so he paused and resumed.

"Something else worth noting is that the foramen itself has no separation. If we assume this person lived to the age of at least five, which seems incontrovertible, the weight of its brain pressing on its cerebellum should have squeezed a part of the cerebellum down through the foramen opening. There should have been an aneurysm of significant size."

"How can you know there wasn't?"

"The soft bone would have separated to make room for it. However, this foramen is intact, perfectly normal other than the missing basilar part, which doesn't firmly attach to the skull until some time in the early twenties."

We were nearing the end of the line. He had the air of a man coming out of the zone, ready to move on to something else. "The bottom line, Mr. Pye, is that the brain is too big, has too much downward slope, and not enough internal support, yet it lived to be five or so. I find it extremely difficult to reconcile those disparate characteristics with that obvious fact. I can only explain it by suggesting it had a brain composed of stronger matter than is usual—or it didn't have a cerebellum as we understand cerebellums."

That brought another echo of Colonel Corso's words about the Roswell aliens: *"The medical examiners believed that the alien brain, well oversized in comparison with the human brain and in proportion to the creature's tiny stature, had four distinct sections."* Maybe that was a valid explanation for the Starchild's odd brain . . . *four distinct sections.*

"What about hydrocephaly?" I asked. "Could that explain any of these malformations?"

He shrugged. "No one can rule it out 100 percent because so many variations of it can be expressed. The range of possiblity is infinite. On the other hand, this skull strikes me as far too symmetrical to be hydrocephalic. In addition, hydrocephaly can't account for a shifted and unexpanded foramen; the missing protuberances, inside and out; the overall thinness of the bone, the missing sinuses, the shallow eye sockets, the crease in the parietals. . . ."

He trailed off, head shaking with apparent bafflement, so I hit him with my best shot. "Would you be interested in conducting a formal study of it? Put it through the scientific wringer and see what comes out the other end?"

He stared at me with his baggy, woeful eyes, now alight with a hotly burning inner fire. "You came to see me about this skull, as I understand it, because of my reputation among my peers. Is that correct?"

I nodded. "I was told you're the best at what you do."

He smiled again, evenly, almost sweetly. "I dearly value that reputation, Mr. Pye. I worked hard to attain it. I'm not prepared to throw it away lightly."

"I'm not asking you to throw away your reputation, Dr. Black. I'm asking you to use it, like a tool, to solve what we both seem to think is a worthwhile mystery."

His baggy eyes narrowed to stare daggers at me. "No! What *you* think is a mystery! I consider it merely puzzling."

I couldn't help spluttering, "But . . . you said—!"

He cut me off with an upraised hand. "You asked me to tell you what I thought about your skulls, which I've done. Now let me tell you what else I think. No matter how odd this skull is—or any other skull, for that matter—it never can be, it never *will* be, considered by science as anything more than an extremely unusual deformity. There is simply no room in our present system for it to be anything else."

"But you know it actually *could* be something else?"

His frosty manner turned frigid. "I know no such thing, Mr. Pye, and if I ever see my name associated with this in any way, shape, or form, you will hear from my lawyer the very next day. Am I making myself absolutely clear to you?"

I rose from my seat to put the skulls back in their box. As I did so, I glanced at Dr. Black and caught him gazing almost longingly at the Starchild. *He knew! He knew. . . .*

I left his office thinking I needed to adjust the odds that the Starchild could be the hybrid offspring of a human and an alien. *Maybe 50 percent hybrid, 50 percent deformity.*

CHAPTER SEVEN

LINCOLN, NEW ORLEANS

Acceptance without proof is the fundamental characteristic of western religion. Rejection without proof is the fundamental characteristic of western science.

—*Gary Zukov*

I spoke in Denver, then again in Crestone, before heading home to New Orleans by way of Arkansas and a speech in Mountain Home. Along the way I stopped to visit a friend in Lincoln, Nebraska, Scott Colborn. Scott owned a metaphysical bookstore and shop, and one of his friends introduced me to Dr. Fred Mausolf, an ophthalmologist specializing in, among other things, eyelid surgery. He had written a book on eye orbits, *Anatomy of the Ocular Adenexa*. I figured if anyone could tell me about the strange eye sockets of the Starchild skull, it would be Dr. Fred Mausolf.

He asked me to bring the skulls to his home so he could have uninterrupted time to study them. He lived in an upscale part of Lincoln, and he and his wife welcomed me warmly. He was another man in my age range, as gracious and forthcoming as Dr. Black, but of average height and solidly built. Nearly invisible rimless glasses caught my attention because they were just becoming fashionable. After the usual chitchat, in which I informed him of my father's forty-year career as an optometrist, we settled ourselves at the living room table and I pulled the skulls from their box.

I told him what I knew about how they were found, saying nothing about what any other specialists had said. As with the psychics, I tried to keep one from influencing others. To his credit, Dr. Mausolf didn't jump to any rash conclusions that would later force him to find a graceful way out of foot-in-mouth disease. He held it in his hands and studied it, quietly, deliberately, focusing on its many anomalous details.

"To just look at it," he began, "the orbital fissures are quite different. Notice the roughly L-shape they make in a typical eye socket." He nodded toward the normal skull. "Now look at this one. It has a rather enlarged superior orbital fissure, while its inferior fissure is truncated to the point of loss, here." He pointed to it with a pen tip, and I could see he was right.

"Is that unusual?"

"Very much so."

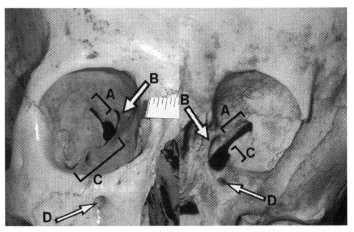

A comparison of eye sockets between SFWS (left) and the SC (right). Alignment along the upper nasal bones gives the correct locations of the respective orbits. A – superior orbital fissure, B – optic foramen, C – inferior orbital fissure, D – infraorbital foramen.

He left it at that and shifted his focus to start rubbing his right forefinger across the inner surfaces of both sockets, feeling them as if he were reading braille. His long silence got to me, making me feel I should offer a bit more information.

"Someone suggested it might not have eyes like ours," I said, leaving out that it was the consensus opinion of the UFO Congress' board of directors. "They felt that if such shallow sockets contained eyeballs like ours, they'd bulge out like frog eyes, which meant they couldn't to be fully protected. If one was so exposed, even a slight bump might tear it right out of its socket. That doesn't seem like a practical design."

Still rubbing his forefinger tip inside and all around both sockets, Dr. Mausolf barely nodded at my comment. Like Dr. Black's probing finger inside the foramen magnum, his kept stroking the smooth inner surfaces of both sockets.

Stroke . . . think . . . *stroke* . . . think.

When he finally stopped, he looked over at me to ask, "What kind of eyes did they think it might have?"

"Dome-like," I replied, making a curved motion with my hand. "Imagine an insect's eyes. Hundreds, thousands of identical receptors fused together into a functioning unit. Put eyes like that on it and it would have 180 degrees of vision, or more, so the neck wouldn't need to move much, either."

Dr. Mausolf smiled indulgently. "How could a human being—even a badly deformed one—develop an *insect's* eyes?"

I shrugged. "It was only a suggestion by someone not as qualified as you. That's why I'm here talking to you."

He handed the skull over to me. "Look at the surfaces of the eye sockets. See anything unusual? Any convolutions?"

I gazed at them for the hundredth time. "No."

I'd had the skulls for nearly three weeks, yet still hadn't thought to stick my fingers in the eye sockets. I had a built-in caution from childhood—*Be careful! Don't put an eye out!*—but they also looked perfectly smooth.

"Feel around in them," he said. "Tell me what you find."

I did as he instructed, running the tip of my forefinger around the inside surface of both sockets. What you could not see with your eyes—and mine were still pretty good, even without glasses—was quickly picked up by sensitive nerve endings. "Very shallow hills and dales, ups and downs."

"Good. Now put your fingertip on the hollow there. . . ." he said, pointing to a spot on the upper outside corner of the left eye socket, at two o'clock as I looked down at it. "Now put your other finger over there." He indicated an opposite area on the right socket, ten o'clock as I looked at it.

I felt two small, very subtle concavities in the surfaces.

"Drag both fingers down and across, to the foramens."

My fingertips slid over a slight rise in the contours, sinking back into another subtle concavity at four o'clock in the right eye socket, eight o'clock in the left. They were *exactly* alike.

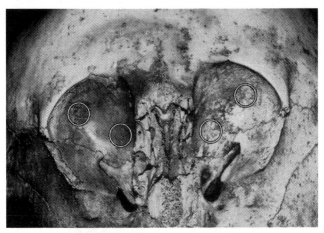

Precise shifts of terrain in SC's eye sockets are marked by circles.

"Those hollows in both sockets correspond to attachment points for muscles that control vertical eye movement. I'm not saying this is a typical human eye socket. Far from it. But it's not an insect's eye structure, either."

"Okay, scratch the insect idea. What kind of eye is this?"

He pointed to the upper border of each socket. "Notice how both of these come to an edge. None of the rounding in a typical eye socket. So the eyeball itself must rest flush against it, probably the entire way around the sockets. That could, as you mentioned, give it a bulging, frog-like effect."

I nodded, pleased to hear his support for our first guess.

"Also," he added, "the forehead extends down in a flat plane, nothing like the rounded protuberance we usually find."

"It has no brow ridge," I said, already aware of that fact.

"None, plus no dip between nose and brow. Not unheard of in humans, but not anything we see on a regular basis."

I nodded again. This was getting good.

"To me, though," he went on, "the upper eyelids are the most unusual part of the anatomy. Assume normal eyes were there. Somehow, extra large eyelids had to connect into this greatly reduced brow, and on every blink they had to extend out around the eyeball to no less than the midpoint of the nose bridge. Finally, they had to fold back somewhere along that sharp upper socket ridge. The question is, how did that much skin tuck itself into such a relatively cramped space?"

Bust of a Grey created in 1978 at the direction of an abductee with a distinct memory of his abduction. Note the extra long and unusually shaped eyelids, with no obvious eyelid crease. (Image used courtesy of Wendelle Stevens.)

"Do you consider it impossible?"

"No!" he almost yelped. "Nothing like this is impossible. It's just an unusual combination of effects, and if you try to imagine how it might have looked, it's difficult to reconcile."

"My main question is, how could standard human eyeballs fit in, and function properly in, such shallow sockets?"

He smoothly adjusted his barely visible glasses. "As we agreed, if normal eyes were in place, they'd probably bulge off the face too far to survive the rigors of a typical childhood."

"So what are you saying?"

"It could have been born blind—no eyeballs at all."

Blind! Somehow that obvious possibility had never occurred to me. "But if it was born blind, why would the inner aspects of the sockets be so perfect?"

I had already noticed that the foramens and fissures—the openings for the optic nerve, and veins and nerves attached to the eyeballs—were precisely aligned in both sockets. Now I realized the surface convolutions were even more precise. With so much precision in the alignments and convolutions, how could they result from deformity? Typical growth almost never produced such a high degree of symmetry.

Dr. Mausolf smiled at my question. "That's what puzzles me. How does such precise symmetry result from what is clearly a deformity? I have no answer for that. Maybe we have to consider it, 'just one of those things'."

Maybe we do.

I arrived home to find the usual barrage of emails stacked up, but the Starchild had become my front-burner issue. I made some calls and was directed to Dr. Joseph Smith, head of radiology at Children's Hospital of New Orleans. With hardly a question he agreed to meet me. "Unusual skull, eh? Sure, bring it on over." No hemming or hawing with him. Just by talking on the phone, I liked him.

He was a tall, looming man with a genial smile and the easy manner required of someone who spent a career working with children in distress. He showed sincere interest in everything around him, from a ten-year-old's badly skinned elbow to the unusual skull presented to him by a stranger.

First he showed it to every person on his staff, people little different from Melanie Young in El Paso. Like her, they tried to call to mind a disorder covering its complex array of anomalies. Like her, they could only conclude that if it was, indeed, anything natural, it wasn't caused by a known genetic deformity. It had to be a birth defect caused by a sperm-ovum misconnection. Because they worked in a children's hospital, they spent much of their time coping with such one-time-only deformities, especially ones that created disastrous results.

When Dr. Smith had heard every suggestion anyone wanted to offer, he said, "In a general sense, it doesn't express the hallmarks of a typical birth-defect deformity. No apparent fusion in the cranial sutures, no asymmetry. But we really should give it a CAT scan to see what shakes out."

"Ummm. . . ." I stammered. "How much will that cost?"

He dismissed my concern with a knowing smile. "This one's on the house—I'm curious about it."

In ten minutes he had arranged for, and executed, a beautiful series of CAT scans of the internal structure of the skull, starting from the top of the head and going down to the neck opening in a series of slices with an electronic "knife."

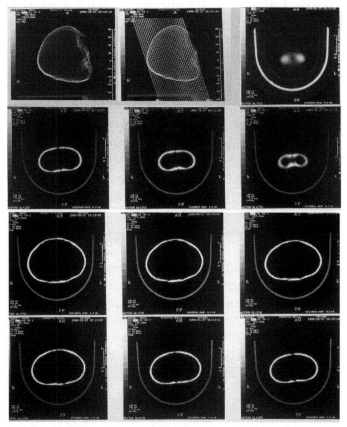

CAT scan images of Starchild skull.

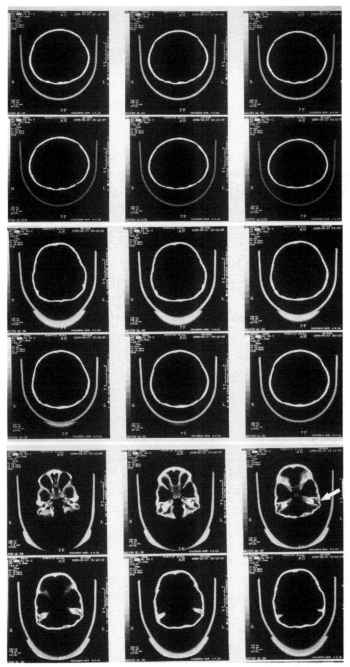

CAT scan images of SC (continued). Arrow at far right points to inner ear region.

"See these sutures?" he said, matter-of-factly. "Those little openings where the edges of the skull plates meet tells us that each one is healthy. There's no fusion anywhere, which rules out that kind of genetic or congenital deformity."

"Does it rule anything *in*?"

He grimaced. "The range is wide open. Just because we've never seen anything like it doesn't mean there's no answer."

"Could you speculate a bit for me?"

He turned back to the rack of CAT scan sheets, hung up against a backlit viewer. "You already know some of what we see here. Uniformly thin bone throughout, top to bottom. That's highly unusual. With deformity of any kind we expect variation between affected and unaffected areas. Its sinuses are missing. Not impossible but exceptionally rare. In over forty years of doing this, I've never seen zip there, nothing. Highly unusual also. And the inner ears . . . those must be twice the usual size. I can't even imagine what that means."

He paused to consider, then turned to me. "Let me make a guess here, take a stab in the dark. I think it's possible this individual was born with a congenital defect that rendered it physically helpless. It had to be cared for its entire life, and it spent that life, such as it was, lying flat on its back, which would account for the extensive flattening in the rear."

"Would it?" I asked. "My understanding is that the attachment of the neck muscles to the inion prevents any flattening past that point, yet this skull is flat well below the inion. In fact, there's *no* inion at all, just a shallow depression. And the neck is extra thin, maybe only half as thick as average."

Dr. Smith smiled again, one of many I saw from him that day. "It's not a perfect theory, just the best I can come up with based on what we've learned from observing it and scanning it. I wouldn't take it to the bank."

"Do you see any other possibility? Anything at all?"

He winked knowingly. "Nothing I'd be inclined to put my name to, but you're free to speculate all you want. I'll tell anyone who asks me that I think it's a damned weird skull sample. Anyone can see that."

Yes, anyone could.

After returning home, I started a practice that in retrospect is laughable, but at the time was deadly serious to me. The paranoid element in the UFO community had convinced me that our government would ultimately seize the Starchild. The only question was, what would be done to me in the process? Would I be left alive and thus able to tell the tale of what happened to me? Or would I be rubbed out, Mafia style, to teach a harsh lesson to those inclined to push too hard on the *verboten* topic of alien reality? With those concerns firmly in mind, for several weeks I went to sleep every night worried that the next morning I might wake up dead.

To soothe my frayed nerves, I called on my military intelligence training to offer a "soft" target to anyone assigned to confiscate the skulls. Across the street from my apartment complex was a shopping mall with a few hundred places to park cars. At noon each day I moved my car from my complex's parking space to the mall, leaving it for an hour or two with the skull box sitting in the back seat. All anyone had to do was bash in a window and be gone. I didn't worry about ordinary thieves stealing it because a lone cardboard box wouldn't be worth the risk of an in-the-open smash-and-grab. But to someone studying my routine, and who knew what I kept in the box, it would be an ideal no-danger, no-testimony heist. How could they resist?

After a month of doing that, I couldn't maintain my level of anxiety about it. I started just leaving the box in my car in my parking space. After two weeks of that, I brought the box in my apartment and kept it there from then on. If they hadn't made their move in six weeks, they weren't going to.

Based on my experience, rumors of no-questions-asked confiscations of UFO-related items might be somewhat overstated. I'm not saying they're impossible, or that they never, ever, no-matter-what don't occur. I can, however, state unequivocally that I saw no signs of it. My phone may have been bugged, my emails may have been read, they could have followed me everywhere, but no one ever openly bothered me.

I still get emails from people who've heard the rumors and ask me if I've ever been harassed in any way. When I tell them the truth, they can never disguise their disappointment. One

Texan summed up those reactions with this neat quip: "Then you ain't got nuthin' they want, pod'nah. If yer skull was a real alien, they'd of already took it from you."

"Maybe they don't believe it's a real alien," I replied. "Maybe it's a guessing game and they're guessing it's a deformity."

"Yeah," he drawled, "or maybe they don't think you can do much with it even if it *is* real. Ya ever thought o' that?"

Many times, my friend . . . many, many times.

I attended Tulane University in the mid-1960s, where I took a basic anthropology class. It was a good department then, and because I still received the university's quarterly magazine, I knew it had gone on to become one of the most highly regarded in the U.S. for its comprehensive analysis of the ancient cultures of South America—especially the Aztec, Incan, and Mayan cultures. Consequently, I figured that if anyone could provide dependable information about the Starchild skull from a topnotch academic perspective, it would be the anthropology department at Tulane.

I called the department's switchboard, explained that I had an unusual skull that needed analysis, and was directed to Dr. John Verano, a forensic anthropologist specializing in the physical analysis of ancient bones. He sounded perfect.

Like Roger White at the University of Nevada at Las Vegas, Dr. Verano grilled me on the phone to determine how I possessed an unusual human skull in the first place, followed by why I felt I needed his help to evaluate it. In the end, I was able to convince him I only wanted an understanding of it that made sense, that I could believe—nothing more or less.

Tulane is an upscale private university in the heart of an upscale residential district in New Orleans. You couldn't tell that, though, from the building that housed its renowned anthropology department. It was an old frame house along a part of Fraternity Row. I was surprised and disappointed at this discovery. They seemed to deserve much better.

We met in Dr. Verano's office. He turned out to be a decade or more younger than me, about my size, but wiry and intense. He informed me that one of his colleagues would be joining us. Dr. Trent Holliday was younger by another decade, already a recognized specialist in Neanderthals. He was more relaxed and outwardly friendlier than John Verano's straightforward, no-nonsense manner.

John's office was the typical academic's jumble of books, papers, and—because of his specialty—scattered artifacts and fossils. I showed them the skulls and was surprised by their quick response. John held the Starchild in his hands no more than a minute before speaking.

"There's nothing really unusual about this skull," he said. "It's rare, I'll grant that, but we find ones like it regularly enough throughout South America."

Naturally, I was disappointed to hear that, but from the beginning I'd been waiting for this kind of shoe to drop. In my heart of hearts I could see no conceivable way a human-alien hybrid skull could have found its way into my care. It would be like that shepherd finding the Dead Sea Scrolls.

"What is it called?" I asked. "The name of the disorder?"

"Cradleboarded hydrocephaly," he calmly announced.

Whoa! What have we here?

I began trying to explain why I couldn't agree with that assessment, but before I could get rolling John was shuffling through photos of some of the thousands of skulls he'd dealt with in South America. After a while he found what he wanted and laid out half-a-dozen on his desk. Most presented a front and a side view, similar to mug shots.

"You see," he said, "the eye sockets are the same."

I had to admit, the eye sockets in these photos were indeed shallower than usual, but they weren't as shallow as the Starchild's. Also, the rears of their heads were flattened by cradleboarding, but not past the inion, as was the case with the Starchild. That area had only a slight resemblance.

"The eye sockets are shallow enough to be comparable," I admitted, "but what about the rears of those heads? This skull is totally different."

Trent brushed that off. "Human skulls show extremely wide variation. You never get matches in all reference points."

"You're lucky to get a few," John added, "and these unusual sockets and the occipital flattening are enough to draw acceptable parallels."

"With all due respect," I replied, "I don't think a case like this should be judged the same way you usually do it."

"What's so special about this case?" Trent asked.

I knew it would come to this eventually, so I went ahead and confessed. "We think it may be a human-alien hybrid."

Their expressions of stunned surprise can be easily envisioned, but because they were trained to be gentlemen at all times, they didn't burst out laughing. However, it was all they could do to keep from smirking at me.

"You're serious?" John asked, struggling to sound serious. Trent glanced sideways, trying to avoid making eye contact with his friend so he wouldn't guffaw in my face.

I nodded, firmly but nowhere near confidently. "That's our working hypothesis until someone can provide solid reasons to think we're barking up the wrong tree."

John tapped the photos on his desk. "These are solid."

I shook my head. "Not for me. I need a closer match."

They glanced at each other, still struggling not to laugh out loud, which I've always admired about both of them. The whole thing could have gone so much worse.

John rose and Trent followed him up. "We've told you what we think. It's up to you to use it however you see fit."

"Can I have permission to use your names in regard to this meeting? Will you vouch publicly for what you've said?"

They exchanged glances, then John spoke. "I don't see why not. But if anyone asks us about it, we'll tell them we think it's a typically deformed human, nothing more or less. Certainly not a human-alien hybrid."

"All right, fine. Thanks for your time."

Both were as good as their word. After that meeting I talked to them a few more times, and each time they were cordial and as respectful as propriety dictated. In fact, John later became my "go-to" guy when journalists or television producers would ask me to recommend someone to publicly give a view in opposition to mine. I'd always tell them, "Dr. John Verano at Tulane," and he'd oblige by giving them his opinions.

Recently I contacted John to offer him a chance to formalize his opinion in this book. This is his considerate reply:

"Thank you for the invitation to contribute to your book. I always appreciate your openness to contrary opinions. Unfortunately, I must pass on this because I'm already behind on too many promised articles, and my examination of your skulls was very brief. You are welcome to cite my opinion that both the adult and child show what I think is a common cultural modification (cranial deformation), and that I don't find their morphology to be abnormal. But I don't have more to say than this, really, based on what I could see from my visual examination. I wish you success with your book project, and on your continuing research regarding these skulls."

With this much under my belt, I knew I needed to educate myself regarding the ins and outs of human physical deformity. In New Orleans, I had access to medical schools of both Tulane and Louisiana State University (LSU), so a wealth of technical information was within easy reach. After a week in the stacks of each school's medical library, I knew enough to state with confidence that it was highly unlikely the Starchild skull resulted from a recognized deformity. I didn't know what it was, but I knew what it wasn't.

Deformity, it turned out, came in two broad categories: genetic and congenital. Genetic deformity came from flaws in the human gene pool that could appear at any time, in any place, and resemble their expression at any other time or place. Included in this category were the often-mentioned hydrocephaly (water on the brain); Down's syndrome (called Trisomy 21 for the extra 21st chromosome that initiates it); Trisomies 13 and 18 (also caused by extra chromosomes); Crouzon's syndrome (an early closure of the sagittal suture); Apert's syndrome (Crouzon's more serious form); Progeria (severely premature aging); and Treacher-Collins syndrome.

Expressions of these disorders range from unflattering to repugnant, so I'll spare readers from enumerating details. For anyone wanting to know more about them, information is readily available in libraries and on the Internet. The point was—and remains—that no single disorder could account for the myriad variations the Starchild skull presented.

This brings us to congenital birth defects, which are the

serious genetic flaws that occur most often at the moment of conception. These were the highest hurdles to clear when trying to convince reluctant scientists to stretch their imaginations to consider the Starchild as potentially something other than entirely human. Almost without fail, they'd parrot their party line of cliché: "When it comes to sperm-egg misconnections, *anything* is possible. That means *anything* anyone puts before us, no matter how bizarre or misshapen, is *natural* and therefore perfectly *normal,* no matter how *ab*normal or *un*natural it might appear to the untrained eye."

I've heard more variations of this than I care to remember, up to and including from no less a luminary than the late Stephen J. Gould, after I asked him to grant the Starchild skull the honor of his widely esteemed consideration. With him and nearly every other credentialed "expert" I consulted, the hardest part became accepting that while those I dealt with might have considerable knowledge about skulls in general, none knew more about the Starchild skull than I did.

I had taken the time to study it, which they wouldn't bother doing because they didn't feel it needed more than a cursory examination. In general, they believed they already knew all they needed to know about every aspect of their bailiwicks, which made it difficult to confront and deal with something truly anomalous like the Starchild. They could only fall back on the mantra they had been taught and wholeheartedly believed: that there could never be anything truly extraordinary or unique anywhere within their accepted purview.

Bertrand Russell summed it up best for all researchers who challenge accepted paradigms: *"What men want is not knowledge, but certainty."*

At the same time I studied the finer points of deformity, I also brushed up on hybridization, which I had to learn quite a bit about in the writing of my book, *Everything You Know Is Wrong.* In Part IV, I discussed the writings of Zecharia Sitchin, a Sumerologist and ancient history theorist who was well-known and widely respected in the world of altenative knowledge. His books dealt mostly with the writings left to posterity by the ancient Sumerians. His work was truly groundbreaking, and therefore was highly controversial because human hybridization at the hands of alien "gods" was at its core.

The Sumerians left to posterity as many as 100,000 tablets written on soft clay in a language known as cuneiform. After completion, those tablets would be fired in a kiln and turned to stone, thus becoming the benchmark for veracity in ancient times—"written in stone." Most of the tablets detailed everyday workings of Sumerian society and culture, which, inexplicably, sprang out of nowhere about 5,000 years ago to achieve an extremely high level of sophistication overnight in historical terms. In practical terms, it seemed impossible.

The Sumerians produced over 100 of the "firsts" we now attribute to a high civilization, though writing is the only one they are consistently given credit for. They were, by all accounts, a civilization so "sudden" that they impress even the most hidebound establishment scholars. However, what makes the Sumerians so special in history, and in the writings of Sitchin, myself, and others, is the written record they left regarding what the establishment calls their "mythology," but what alternative scholars call their "history."

Part of their history, according to alternative scholars, was somehow knowing that Uranus, Neptune, and Pluto were in our solar system, 4,000 years before we rediscovered that lost knowledge. They didn't label Earth as the "third rock from the Sun," as we do; they regarded it as number seven because they started with Pluto and counted inward. (And isn't *that* an unusual perspective for a people supposedly not long out of caves and just past hardscrabble subsistence farming?) They also included a currently "missing" planet in their scheme of things, a very unusual one they called "Nibiru," on which lived their "gods," the Anunnaki.

According to the Sumerians, the Anunnaki could travel to and from Nibiru in spacecraft, and 600 of them started settling on Earth around 400,000 years ago. By around 300,000 years ago, the transplants decided to create for themselves (in their "house of fashioning") new plants and animals to "give the gods their ease" (make life on the distant outpost more comfortable and, presumably, more like home). They also decided to create a "slave and servant" to further make their life on the outpost as comfortable as it could be.

From this came our modern domesticated plants and animals, among which were the "Adamu," the first humans. (The term "Adamu," the slaves and servants created by the Anunnaki "gods" of the Sumerians, was slightly modified 2000 years later to become "Adam" in the Old Testament.) The Anunnaki created the Adamu by hybridizing their own "essence" with the essence of "creatures of Earth," the planet's indigenous upright walking beings. We assume those were either Homo Erectus or Neanderthals, both of whom were in existence when the Adamu were being created.

The "creatures of Earth" provided an appropriate physical model for a functional slave and servant formed "in our own image, after our own likeness" (another set of words on a Sumerian tablet copied 2,000 years later in Genesis). The new Adamu would be smart, but not too smart; strong, but not too strong—i.e., useful but not inherently dangerous.

In this hybridization between the Anunnaki (the "gods" who lived on Nibiru) and the "creatures of Earth," I heard echoes of what I thought I might be seeing in the Starchild.

Most people have the idea that hybridization is a 50-50 division, half from the parent of one species and half from the other. It never works like that. Even in normal conceptions, one parent will inevitably dominate how you look (your "phenotypic expression"). So the spread can be 60-40 or 70-30 or even 80-20. However, in the case of human beings, if the Sumerians were correct about us being genetically engineered in the Anunnaki "house of fashioning," then the split in our phenotypic expression could be 90-10 or even 95-5.

There can be little doubt that, apart from our bipedality, humans have virtually nothing in common with Homo Erectus or Neanderthals. Then, in a widely reported study in 1987, our mitochondrial DNA revealed that our genetic history as a unique species, as the beings we are today, extended back to only about 200,000 years, *precisely when the Sumerian tablets indicated that the Adamu were created!*

Humans, it turns out, have 46 chromosomes, while all other higher primates have 48. How could two entire chromosomes go missing from our gene pool, yet we are so much more mentally sophisticated, so much "better," than higher primates? To me, something was absurdly dissonant about that fact, and in my studies I found out what geneticists know but seldom advertise, so no one of any importance will notice: *Those two chromosomes aren't missing!*

By a seeming "miracle" of genetic good fortune, somehow humanity's second chromosome is actually a *fusion* of the second and third chromosomes of higher primates! So we keep all of the higher primate gene code in those two chromosomes, but we have it massively rearranged from what Mother Nature could rationally be expected to produce by an accumulation of miniscule mutations.

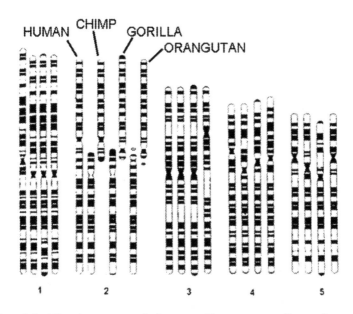

Depiction of first five chromosomes in humans, chimpanzees, gorillas, and orangutans. Note second and third human chromsomes are fused into one. (An adaptation based on research published in "The Origin Of Man: A Chromosomal Pictorial Legacy," a 1982 article in Science, Vol 215, Issue 4539, 1525-1530 by J. J. Yunis and O. Prakash.)

The fusion of our second and third chromosomes simply could not occur by mutations alone—it required something very much like *intervention,* by brains and hands capable of manipulating it in . . . well, a "house of fashioning" sounds as if it could have been much like a genetics lab, doesn't it?

Also, look carefully at the pinched areas along the upper parts of the fifth chromosome. Notice that the human and the orangutan are very similar, the gorilla is very different from the others, and the chimp's has somehow been flipped in the area around the pinch, turned upside down around the axis of the pinch. How would nature manage that? And why?

There is much other valid evidence to support what is becoming known as the "Intervention Theory" of human origins. But what it meant to the Starchild case was that it had given me a firm grasp on how a hybrid being might be created between a human and a Grey alien. On the surface, they would not seem able to procreate, almost certainly not by a normal sexual union. However, if the union could be genetically engineered, then virtually anything became possible.

You would have to start with the egg of a human female, because even with the Anunnaki, it began with female eggs. Why? Because the size of eggs relative to sperm make them easier to work with in a genetics laboratory. You have to utilize both to make a hybrid, but the sperm contains only its package of chromosomes and genes, while eggs contain all of that *plus* all of the mechanisms needed to carry a fetus through its gestation to birth. So, if you intend to hybridize between a human and a Grey alien, you should start with a human mother. Once that condition is met, you can skew the balance of traits as much as your genetic expertise permits.

This is how you get human beings looking absolutely unlike any other higher primate on the planet. We seem to have indeed been made "in the image and after the likeness" of the "gods" who created us, gods not fully adapted to our planet or its gravity. This is why in many ways we are physically maladapted to our terrestrial home, as evidenced by our sunlight sensitive eyes and skin, bad lower backs, varicose veins, hemorrhoids, and similar gravity-induced debilitations.

The point is that the Starchild could indeed have been a hybrid between a human mother (herself from a hybrid species created 200,000 years ago) and a Grey alien father, and it could be a perfectly viable offspring while being 90 or 95 percent like its father and only slightly like its mother. This was the only way I could explain the stunning lack of human corollaries for the features in the Starchild skull. Virtually *everything* about it was unlike any human counterpart, yet it had lived and functioned at least until the age of five.

Without my previous understanding of hybrids, and how they could be genetically created by those who knew what they were doing, I could never have pressed the Starchild's case with the confidence I was beginning to feel about it.

Once I had convinced myself I wasn't on the government's hit list, I went to Wal-Mart and bought a durable, lockable tool box to hold both skulls in a foam bed cut to accept them. Then, at a dumpster at my apartment comlex, I held a brief ceremony of respect for duty well served by the El Paso cardboard box. I never knew how many years it held its prizes.

That serious-looking tool box (gray, of course, in honor of its contents) added a decided uptick to my credibility as I continued taking the skulls to scientists or doctors who agreed to examine them. In addition, I took them to psychics and other "sensitives" who wanted to "psychometrize" them. I wasn't picky and I didn't play favorites. I wanted and needed solid, verifiable, provable information, and I didn't care where it came from as long as I felt I could count on it. Thus, the bottom line gradually, surprisingly revealed itself.

Scientific "experts" knew no more about what they told me than psychics. And while we're at it, toss in UFOlogists. Between those widely disparate groups, I couldn't see any difference in insight or understanding. Whatever their outward intentions, all three groups did little more than guess, just tossing mental darts at the problem, hoping to hit somewhere near it. The most consistent thing was that everyone I talked to—scientist, psychic, or UFO buff—sounded convinced they were correct. What none of them realized was that now I knew enough to know they were guessing.

All they did was leave me wondering what to try next.

CHAPTER EIGHT

DOING THE JOB

New opinions are always suspected, and usually opposed, without any other reason but because they are not already common.
—*John Locke*

I don't want to paint too dark a picture of those hectic months in 1999 when I struggled to figure out exactly what I was dealing with. In mid-April I was a guest on the Art Bell radio show, *Coast to Coast AM*, when he was riding high and commanded a late-night audience in the millions. I discussed the skull in considerable detail over the course of three hours, and urged people to visit our new website. Tens of thousands of hits were tallied over the next several weeks, which resulted in a wide range of relevant emails.

A Minnesota dentist named Dr. Jary Larson offered to take our case to a friend, Dr. Robert Gorlin, of the School of Dentistry at the University of Minnesota. Dr. Gorlin, he wrote me, was "a world famous authority on syndromes. He has devoted his life to the study of physical and structural anomalies of the human body." Furthermore, Dr. Larson added, Dr. Gorlin "has a DDS, an MS, a PhD, and an MD to go with many prestigious awards and honorary degrees from a host of universities and colleges." This was just the kind of expert we needed, so I got right back to Dr. Larson.

By then our website contained highly detailed descriptions of every significant aspect of the Starchild skull, plus we had posted a number of very clear photos of both skulls, alone and side by side, shot from 360 degrees and from top to bottom. Those images allowed any expert to thoroughly analyze the skulls without holding them. I requested that Dr. Larson ask Dr. Gorlin to evaluate it, and Dr. Gorlin soon replied:

"The skull is a brachycephalic individual with turricephaly, clearly a human who was dysmorphic."

Brachycephalic means "short-headed with a width more than four-fifths the length." Right. The Starchild's width was equal to its length—15 cm in both cases. *Turricephaly* means "caused by premature fusion of the coronal sutures." Wrong. He clearly hadn't considered the CAT-scan results on the website. Turricephaly also produces "a cone-shaped head." What photos did he examine to come to *that* conclusion? The Starchild is odd, but not cone-shaped. And *dysmorphic* is simply malformed or misshapen. Obvious.

I regret having to single out someone like this because many others were equally culpable. However, Dr. Gorlin's response was very typical of what I kept being faced with—one or more solidly credentialed "experts" sounding impressively authoritative, yet actually being well off the mark.

Here I can't fail to mention Dr. Robert Walker, PhD, the head of the Department of Anatomy at New York Chiropractic College in Seneca Falls, New York. After my appearance on the Art Bell show, he promptly fired off this blunt email: "I don't know who your anatomical experts are, but there is absolutely nothing unique about those two skulls. You are simply mistaken at best, a charlatan at worst." Notice the tone of supercilious arrogance, and the mistaken perception that the origins of both skulls were at issue.

I don't want to give the impression that *all* specialists turned their backs and refused to help, or that arrogance or incompetence were all I ever encountered. Let me mention a few people who generously offered their services:

Tom McFarland, the Vermont director for the Mutual UFO

Network (MUFON) connected me to Wes Neville, a forensic artist and sculptor who wanted to work with me on the Starchild. Just before that, however, I was contacted by another forensic artist, Bill McDonald, and gave him the first commission to represent the Starchild graphically. Later, several others created renditions, all of which were dramatically different, making clear how subjective their "art" can be. Of course, with most of its lower face missing, the Starchild skull didn't give those many talented artists much to work with.

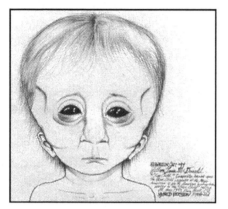 

Bill McDonald's interpretation (left), and Katie Lin's (right), show how polarized such reconstructions can be, especially when forensic evidence is incomplete.

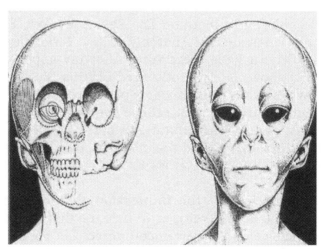

Mark Savee created a compelling interpretation, despite an improbably heavy lower jaw (mandible) and a crown too rounded to be accurate to SC's true shape.

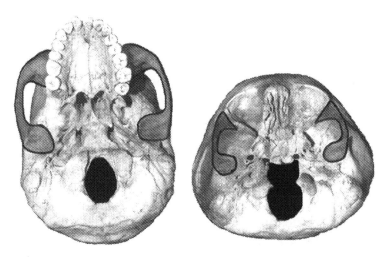

Artistic views of SC are varied because its lower face is missing. Above is a sketched reconstruction of the zygomatic arches (cheek bones) of SWFS (left) and SC (right).

A researcher at the Sims Space Research Center's Department of Physics and Astronomy at the University of Leicester, England, wrote: "As a scientist I believe science should investigate the unusual as well as what might be easily explained. The problem is that most scientists have mortgages, bills, etc., and going out on a limb does not help with contract renewals, grant applications, etc. This includes me, so for the time being I wish to remain distanced from any publicity."

People like Kay Heap, Lyn Taylor, and Jim Frazier offered to do, or to help arrange, work I couldn't afford at the time.

Then Ruth Lowery connected me with Carol Baker, who connected me with Dr. Mohammad Tahir, a forensic DNA specialist who had analyzed bloodstains from the forty-year-old Sam Shepherd murder case. Exchanging emails with Dr. Tahir first made me understand that testing the DNA of the Starchild's aged bone would be absolutely critical, but also quite difficult and prohibitively expensive.

Eventually, I took my tool box's cargo to a lab that was a forerunner of *Crime Scene Investigation*, the now famous *CSI* television franchise. The "Faces Lab" at Louisiana State University in Baton Rouge, Louisiana, was dedicated to putting faces on skulls to try to solve crimes and missing person reports. I showed my odd couple to Mary Manhein, who ran the lab, and to her assistant, Eileen Barrow, both of whom assured me in no uncertain terms that the Starchild *had* to be a deformed human because nothing else they could imagine—or acknowledge to me—was possible.

A retired teacher, Dr. Loys Nunez of Memphis, connected me with Dr. William Rodriguez, a well-known forensic anthropologist at the National Institutes of Health in Washington, D.C. Dr. Rodriguez informed me he would soon attend a conference in New Orleans, so he invited me to show the skulls to him and a few of his fellow attendees, among them Ed Waldrip, the director of the Southern Institute of Forensic Science in Hattiesburg, Mississippi.

Dr. Rodriguez arranged a place for me to show my pair of skulls to a half-dozen forensic specialists, all of whom agreed the Starchild had to be a typical deformity, although there was heated debate about *which* deformity. Dr. Rodriguez and Dr. Waldrip felt it was a cradleboarded hydrocephalic, but Dr. Waldrip later informed me he considered it a rare privilege to have held such an interesting conjunction of deformities in his own hands. He wrote: "For a forensic investigator, it was a once-in-a-lifetime opportunity."

Dr. David M. Harris, a professor of histology at the University of Illinois, offered to do a bone scan if I'd supply a tiny chip from anywhere on the skulls. Inside the adult's nose was an easy area from which to remove a small piece, as this was the same location where the nose of the Starchild had sheared off. It contained several jagged fragments arrayed in multiple layers, so I pried away one of the least conspicuous small ones and, wondering what might be produced, I mailed it and the other chip to Dr. Harris.

A few days later he reported finding a "noticeable deviation" between the two bone samples, but he felt it wasn't enough to raise a red flag. In such cases, he informed me, margins of difference could be exceptionally wide and, in his opinion, these fell within those wide parameters. I accepted that, but asked him to send the scanning slides so I could see for myself what the differences might be.

A friend in the New Orleans medical community connected me with a local pathologist I'll call "Doug," who agreed to walk me through analyzing the bone scans only because our mutual friend intervened on my behalf. Doug was clearly ill at ease with the possibility of viewing alien bone through his twin-sided microscope, through which we could view very thin slices of the Starchild's bone alongside the adult's bone.

Under Doug's microscope I saw two stunningly different views. The adult skull's bone cells resembled hot dogs scattered across a floor, in all positions, at every angle—but all were flat on the floor with open space between most of them. In marked contrast, the Starchild's bone cells were clumped together in wads, jumbled all over the place, like packages of unopened hot dogs dropped onto the same floor, but with *little or no* open space between them. I could make no valid comparison between these left and right views.

(Their thinness made them fragile, so they deteriorated before I thought to have them photographed. That was a mistake on my part, but luckily not a serious one. New slides would be easy to create at any point in the future.)

"These aren't even in the same ballpark," I muttered.

"I have to admit," Doug said uneasily, "that superficially they look like apples and oranges. However, bone scans like these can exhibit huge variations, so such differences are not really beyond the realm of possibility."

"How much farther beyond do they have to be? *C'mon!*"

"Listen," he said, with a bit more edge, "differences like these are unusual, I grant that, but they're not impossible. And even if they were . . . even if this bone was unique in all of history—so what? That doesn't mean your skull is an alien-human hybrid. All it means is that it's a bizarre human."

"No, Doug, *you* listen! It's bizarre in dozens of ways *other* than these scans, so why the hell *can't* it be from an alien?"

Barely controlling his agitation, he said, "There is *no* distance *far* enough from normal to permit a scientist like me to conclude it might be alien bone. We're trained to always look for the simplest, most economical explanation for all anomalies, and the simplest, most economical explanation for what I see here is that it's just a damned weird human—*period!*"

Occam's razor, slicing my way once again.

In mid-August a wicked curveball came at me from Linda Faircloth and her staff at the Pfisterer-Auderer dental lab in New Orleans. She was a cheerful, charming woman my own age, who took her work very seriously. I was sitting beside her at a workbench in the lab area as she examined the detached piece of maxilla with the two molars still in place.

Lowering the magnifying loupe in her hand, she said, "I don't think this is from a child, Lloyd. I mean, the palate is the size of an infant, but I don't see even a six-year-old here. There's just too much wear and tear on the teeth. "

I was flabbergasted. "Everyone else thinks it was five!"

She held the loupe so I could see the teeth magnified. "Notice how the enamel is chipped on the upper surfaces?"

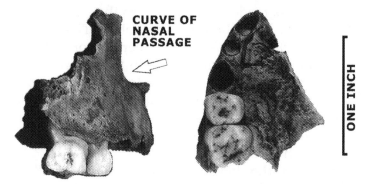

SC maxilla is the size of an infant. While its two remaining teeth could be deciduous (baby), they show signs of considerable wear and tear, which is not typical in children.

At this stunning view, all I could think to do was point to the other skull resting at one side of Linda's work bench. "Its rear molars are flat, so they were stone grinding their corn. Maybe grit from the grindstone caused that chipping."

Her matronly head shook. "Not in a five-year-old. This takes a lot of time and a lot of grinding power. A ten-year-old would have trouble generating as much torque as I'm seeing here. Not only that, the enamel is crazed all around."

"Crazed?" That was how I was starting to feel.

"Crazing is those dark, vertical cracks in the enamel, top

to bottom. The clear ones are from post-mortem shrinkage. They don't count. Focus on the dark ones. You see them?"

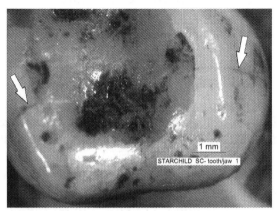

Close-up of SC's small molar showing chipping and pitting on the enamel of the cusps, and "crazing" cracks on the sides (two are indicated by arrows).

I could indeed see both the darks and the clears. "Yes."

"You find it in teeth that get stressed doing things like cracking nuts, and doing a lot of it. The small chewing muscles on this skull couldn't generate enough power."

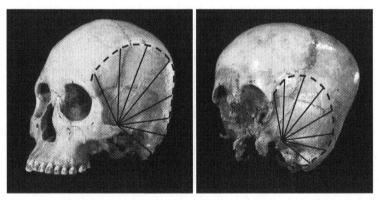

Normal area of chewing muscle attachments in SFWS (left), compared to the considerably reduced area of attachment points in SC (right).

"Okay, assume it was an adult. Why would it have new teeth in position to come down? That doesn't make sense."

"Humans can grow more than two sets of teeth in a lifetime," she replied. "Third-set molars and even incisors can erupt in old age. It's rare, but not unheard of."

This jibed with something Colonel Corso said about the Roswell corpse evaluated by medical examiners at Walter Reed Army Hospital in Bethesda, Maryland. He said the examiners *hypothesized that biological time must have passed very slowly for the entity because it had a very slow metabolism, evidenced, they said, by the enormous capacities of the huge heart and lungs. The physiology of this thing indicated it was not a creature whose body had to work hard to sustain itself.*

"The Bible could be right," Linda went on. "Maybe we used to have extra-long life spans and needed multiple sets of teeth to get through them."

This was the first time I had entertained the notion that a human-alien hybrid might live for hundreds of years and so required shark-like teeth that replaced themselves when one was broken off or otherwise needed repair.

"Another thing I noticed through the loupe," Linda said, lifting the adult to place it beside the Starchild. "Here. . . ." She handed the magnifier over to me. "Study the surfaces of the bone across the top and all around both of them."

This wasn't something I had yet thought to do. Just to look at them, their surfaces seemed pretty much the same. "Okay." After several seconds, I noticed what I had to assume she was talking about. "Those pore-like things are missing."

The surface of the human skull was covered in dark little dimples that looked like pores in normal skin, except much farther apart than pores would be. There were virtually no pores on the Starchild's bone. Some, but relatively isolated.

"Right," she said. "Any idea what those things are?"

"No, but I'll find out."

[Note: They are lacunae, a vital aspect of bone structure and the metabolism of growth. An extreme lack is unusual.]

"Whatever they are, it's strange that they're all over the normal skull but more or less missing from the other one."

"Good point, Linda. I have no idea why they'd be gone."

Neither, as it turned out, did anyone else. Once again, I had to consider it 'just one of those things.'

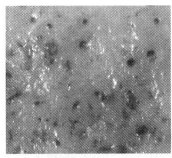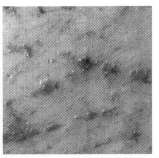

Outer surface of SFWS (left) showing normal lacunae. Compare that to SC's outer surface (right), almost totally lacking in such indentations, which are vital to the growth and subsequent regeneration of bone over the course of a human lifetime.

At the Pfisterer-Auderer dental lab, a team of reconstruction experts normally worked building models of teeth and jaws to create dentures and bridges. Linda Faircloth asked them to make a mold of the piece of maxilla, then make a mirror image of it to create its maxilla. Next, using that, they would try to create a mandible (lower jaw) to align with it.

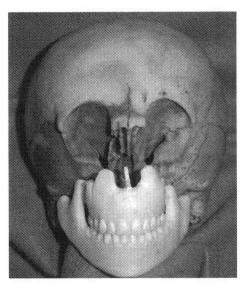

Reconstruction of the lower face (maxilla on top, mandible—the lower jaw—on bottom) created by dental reconstruction specialists at the Pfisterer-Auderer lab in New Orleans.

Unfortunately, all they could use as their model was the human template they knew best, so in the end it came out a bit too wide and thick for what we now imagine the greatly reduced chewing muscles would warrant. Nevertheless, it was enough to begin adding a very rough outline to the mental image we all had of how the Starchild might have looked.

On the less scientific edge of Occam's razor, I was contacted by Josi Galante, an artist in Taos, New Mexico, who was convinced she'd been abducted, impregnated, and her fetus *harvested*, in what is alleged by many UFOlogists to be a program of hybridization between Greys and humans.

[Note: While controversial, there is considerable evidence that supports the claims of women such as Josi Galante. This evidence includes sonogram confirmations of pregnancy, followed by a mysterious "loss" of fetuses in the fourth month.

A few years later, Josi also claimed, she was again taken aboard an alien craft and shown a young boy she was told she could "bond" with because he was her hybrid child. She was convinced the boy was hers, and used her artistic talents to draw him in front and profile views.

Five-year-old boy alleged by his mother to be her hybrid son (courtesy Josi Galante).

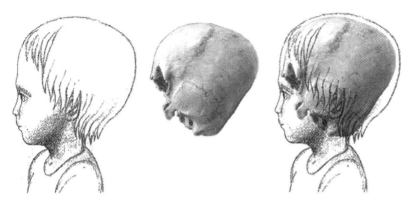

Profile view of hybrid boy drawn by his mother, compared to a profile-view of the Starchild skull. Her drawing was made well before the Starchild came to light.

My memories of him are clear, so I'm confident my details are accurate. Very large eyes, gray with a hint of blue. Hair a whitish to beige or ash color. Complexion very pale, or translucent, almost no blush of underlying pink. These children are all very frail and thin, and are somewhat unkempt and not clean by our standards. The overall description of these little ones is sad, confused, waif-like. [In my opinion] They all seem to suffer from 'failure-to-thrive' syndrome.   —Josie Galante

Failure-to-thrive occurs in any human babies deprived of competent nurturing by parents. They grow into emotionally stunted children who never become well-adjusted adults. Some UFOlogists believe this is why natural mothers are allowed to bond with hybrid offspring, so failure-to-thrive can be lessened. They also feel at least some "missing" children and adults who end up on milk cartons each year are, in fact, *skynapped* to be playmates and caretakers for such hybrids.

While I had the Starchild skull evaluated by a wide range of scientists and psychics, I also brought it to the attention of every media outlet I could think of. Naturally, no one in the mainstream would consider it, but several alternative magazines and newspapers around the world did publish stories about the skull or the effort I was making to get it scientifically tested. In 1999, it was featured in these periodicals:

1. My home newspaper, the *Times-Picayune* of New Orleans, in April scheduled a splash feature on the front page of the Living section, a widely read part. Unfortunately, someone "of influence" (I never learned who) heard about it and called the paper's top echelon to suggest "caution" in featuring such a "volatile" story. After that, the Starchild was buried below the fold (the downside) of the Metro section, one of the least-read parts.

2. Also in April, I traveled to Arkansas to "crash" Lou Farish's Ozark UFO Conference in Eureka Springs, which earned hard feelings from Lou but a full story in the *Northwest Arkansas Times*.

3. In June, I spoke in Telluride, Colorado, prompting coverage in the local newspaper, the *Daily Planet*.

4. *Quest*, an alternative magazine in England, published a story about it in their June/July issue.

5. *UFO Magazine*, an alternative periodical in California, followed that with a piece in October.

6. We hit the "big time," sort of, when *The Sun* (U.S. edition) tabloid newspaper ran one of its brief, breathless articles about the Starchild in October.

7. *FATE Magazine*, a major alternative periodical in the U.S., covered it well in its November issue.

8. Also in November, the tabloid *National Examiner* did a piece. The cover featured Jerry Springer, who attended Tulane when I was there. (We didn't know each other.)

9. A major alternative magazine, *Fortean Times* in England, covered it thoroughly in November.

10. November also saw fair coverage in *Dimanche*, a widely read tabloid type newspaper in France.

11. In December in Japan, *MU Magazine*, a major alternative journal, ran a well-done story about it.

12. Also in December, England's *X-Factor Magazine* covered it for their discriminating audience.

Most of that didn't just happen. It had to be arranged by contacting each magazine and letting them know what I had and asking them to cover it. I didn't contact either of the U.S. scandal rags, *The Sun* and the *National Examiner* (I actually wanted to avoid attracting their attention), or *Dimanche* in France or *MU* in Japan. But the rest were arranged because I asked them to check out the Starchild to see it was a legitimate relic needing to be examined in comprehensive detail. To their credit, they did that and realized I was right.

While all that carried on, I kept up my speaking schedule, which also had to be arranged one-by-one, and then trips had to be made to fulfill the dates. One of my trips was to Los Angeles, where I laid over for a few days with friends from my screenwriting years. While there, I tried to entice several well-known, well-heeled celebrities into taking a look at the Starchild. These were people widely considered to have an interest in any and/or all matters relating to UFOs.

Steven Spielberg, Hollywood's "Mr. Alien," was not interested. Chris Carter, creator of *The X-Files* . . . not interested. Dan Ackroyd, host of a paranormal TV series . . . not interested. Henry Winkler, "The Fonz," a producer of paranormal TV shows . . . not interested. James Cameron, creator of *The Terminator* movies . . . not interested. Arnold Schwarzenegger, the Terminator himself . . . not interested. Richard Dreyfus, star of Spielberg's *Close Encounters of the Third Kind* . . . not interested. Jeff Bridges, star of *Starman* . . . not interested. Jodie Foster, star of *Contact* . . . etc.

You'd think maybe *one* would be just a tiny bit curious, but "their people" assured me they weren't. Nor was Jeff Goldblum (*The Fly*), Sean Penn (iconoclast), Wesley Snipes (open to all things UFO/alien), Shirley MacLaine ("Ms. Woo-Woo" herself, who started talking in a serious way about UFOs and aliens thirty years ago); nor, when I was fed up with the Hollywood shuffle, Barbra Streisand, just for the hell of it.

I also applied to be interviewed on NBC's *The Today Show*, ABC's *Good Morning America*, CBS's *This Morning*, *The Roseanne Show*, *The Rosie O'Donnell Show*, CNN's *Larry King Live*, and *The Tonight Show with Jay Leno*. I didn't bother with David Letterman because I hoped to be taken at least half-seriously. In any case, all of their people assured me they had no interest in something as specious and likely to be a fraud as the skull that I was by then touting as having a real chance of being a human-alien hybrid.

◊ ◊ ◊

TV/radio producer: "It's a weird *what?*"
Me: "A skull. A real, true, bone skull. No doubt about it."
"Who cares about a weird skull? That's not news."
"This one's extra weird. So weird, it may not be human."
"If it's not human, what is it?"
"That's what we're trying to find out."
"Why don't you take it to scientists."
"I've taken it to plenty of scientists."
"What do they say?"
"That it's some kind of deformity, but they're wrong."
"They're scientists! How could they be wrong?"
"They don't know everything about everything."
"If *they* don't know, then who *does?*"
"That's what we're trying to find out."
"Is there a website, or something I can check up on?"
"Go to www.starchildproject.com. It's all right there."
"Did you say . . . *Starchild?* Like . . . an *alien?*"
"More like an alien-human hybrid."
"A *what???*"
"I prefer to call it a human-alien hybrid."
Click!

◊ ◊ ◊

My income stream dried to a trickle as sales of my book, *Everything You Know Is Wrong*, languished from neglect while I focused almost exclusively on the Starchild Project. Donations were earmarked for tests, so I began living off credit cards rather than income, assuming as so many do in similar situations that I'd rally soon and be able to pay it all off in one huge stroke of success. Like any junkie, hubris mired me in credit card debt before I added a load of personal debt by leaning heavily on family and friends.

With the wheels steadily grinding off my wobbling wagon, I needed relief from the burden and it arrived in the nick of time. In late August, I went to the Pacific Northwest to make a round of speeches, and I spent a few days with Chad Deetken and his wife Gwen—witty, charming, hospitable friends I'd made the previous year when speaking in Vancouver, B.C.

Chad is a lean, innocuous-looking fellow with an outward calm that belies the sharp focus he brings to whatever piques his interest. When I met him, he worked for a research organization investigating all manner of paranormal phenomena. His main interests were cattle mutilations and crop circles, but he was flexible enough to make room for the Starchild.

I told Chad I was at the end of my rope and didn't know how to proceed. He suggested we show it to his friend, Dr. Ted Robinson, a craniofacial plastic surgeon. Chad knew Ted had an open mind and would never summarily dismiss us, as most of his colleagues would. I agreed to meet with him, Chad arranged it, and my world changed in a hurry.

Dr. Ted Robinson is an unusual mix of burning curiosity about many of the world's unanswered mysteries, yet he also has firm views about the things he thinks have been sufficiently addressed and answered by science. He was open-minded, but not nearly so much that his brains fell out. He was just the kind of person I'd been seeking for the past six months—a real, true, bona fide specialist willing to consider the Starchild's complete background.

As he listened to me tell it, his steady gaze smoldered with intensity. I was getting to him, making him see possibilities that I saw, and Chad saw, but no other "expert" seemed able to grasp. This skull was potentially *important!*

Ted put the skulls down, leaned back, laced his fingers atop his well-coiffed dark hair, and fixed me with an intense gaze that burned right through me.

Dr. Ted Robinson

"That's a remarkable story, Lloyd, truly remarkable, and I can see it's warranted. In forty years of medical practice, I can't recall ever handling anything like this skull. I can't sit here and tell you for sure that nothing like it has ever been seen before, but nothing I can think of springs to mind."

"As far as I can tell," I replied, "it's one of a kind."

"I'm inclined to agree, eh?" he said, with that rising Canadian inflection on the *eh*? "But I'd like some time to go through the literature relative to deformity. Maybe something like this is hiding in the cracks. For a relic with so much potential significance, we have to be sure about it."

I took his apparent willingness to participate as a clear opening, so I decided to roll the dice again. "If you give it a thorough going-over and find there never has been anything like it, what then? Would you be in a position to do anything about it? Could you take the skull and run with it?"

"Do what you've been doing? Take it to other specialists? Ask for their expertise? Maybe arrange for some analysis?"

"Right! With your reputation and credentials, that should be a piece of cake for you. What do you say?"

We sat in silence for a while as he let it all sink in.

"Give me a month to research it, to see what I can come up with," he finally said. "If I find out it *is* unique, truly one-of-a-kind, then okay, I'll keep it here for a while. I don't think we'll be as rigid in our views as our colleagues in the States. We tend to be more open to things a bit off-center."

"Great! I can't tell you what a relief it is to hear that!"

"And Chad . . . I'd like him to partner with me if he will."

"No problem," I said confidently. "This will be right down his alley." And it was. He readily agreed to join with Ted, and to support him in any way their efforts required.

As September passed, I continued my routine of traveling and speaking to try to bring the Starchild's case to media attention. I had a steady stream of successes with radio shows and print coverage, but still no offers of the technical and financial help I needed. Then Ted Robinson called.

"I think you're right about the Starchild, Lloyd. I went through every reference book in Vancouver, checked everything I'm aware of, and I can't find anything resembling the skull. I now agree with you that it's totally unique, so I'm willing to accept that it might not be fully human. I'm prepared to work here in Canada to try to prove it definitively."

This was what I'd been hoping to hear from a scientist—*any* kind of specialist, really—since I started on the journey seven months earlier. It was now late in 1999, so I had grown eager to hand the Starchild over to anyone. Having that be Ted and Chad was an immense relief. Naturally, I felt guilty for failing, but I was delighted to have this grinding burden lifted off my shoulders and safely tucked under the wing of someone with Ted's credentialed firepower.

"I've discussed it with a few colleagues," he went on, "and I must admit I'm disappointed by their reactions. Not in every case, but for the most part they convinced themselves it has to be a congenital deformity. I know you said it would be like that, but I thought you were exaggerating."

"Yeah, well, the general close-mindedness out there surprised me too when I first encountered it. Now I find it easy to just ignore people like that. They speak from total ignorance of knowledge they have no desire to learn."

"I'll try to keep that in mind. Now, on another front, a friend of mine, Dr. David Sweet, runs a new forensic dental lab at the University of British Columbia here in Vancouver. He thinks he may be able to recover DNA from both skulls. That means cutting into them, eh? But you have to start thinking in those terms. Eventually, DNA will be *the* critical result you'll have to secure to make your case stick."

He didn't have to tell me that.

After six months of talking to experts and reading relevant material, the bottom line was that only a comprehensive analysis of the Starchild's DNA could establish its parentage. That problem was exacerbated in mid-summer, only a few weeks before meeting Ted, when I finally arranged (and could pay for) a Carbon 14 test for the adult skull at the University of California at Riverside. I didn't bother dating the Starchild skull because I had barely enough money for one, and I assumed both skulls died together and were found in the manner I was told. The C-14 result showed the adult skull was 900 years old, plus or minus 40 years.

Relatively inexpensive DNA testing could be done for a maximum period dating back to 50 years. Anything older was considered "ancient" DNA, which required very special handling and a corresponding rise in cost. Thus, the DNA we needed to recover from the Starchild wasn't merely ancient, it was *seriously* ancient, and would be expensive to recover.

On the upside, if the skulls had indeed been inside a mine tunnel all that time—out of wind, rain, sunlight and airborne bacteria—the likeliest result seemed to be a typical degree of desiccation brought on by lengthy burial in a dry climate. That left me optimistic about our chances for DNA recovery— *if* I could find a lab with scientists willing to risk their reputations by doing what their high priests forbade.

Unfortunately, recovering ancient DNA was such an exacting process that worldwide only a handful of labs were equipped to do it, although new ones were coming online in a dozen or more countries as the expensive equipment and extensive air filtration systems were secured and installed. I had already been rejected by the handful now operational. Among them was the Max Planck Institute in Leipzig, Germany, headed by Svante Paabo, the world's foremost analyst of ancient DNA, who glamorized his field by recovering mito-chondrial DNA from a 30,000-year-old Neanderthal fossil. He showed their genetic relationship to humans wasn't close enough to put them on our direct evolutionary line.

Three days before Ted Robinson's call, I found out my top lab prospect in the U.S. was afraid to become involved. Their rejection said in part, *Ours is a young company, and it is bad publicity to associate with unprofessional projects based on such flawed ideas.* So Ted's news about his friend David Sweet's new genetics lab was especially welcome.

"Wow!" was all I could think to say. "What great luck!"

"You have to come to Canada to meet David. Melanie and Ray should come, too. Everyone needs to be on the same page before he does any cutting and testing. Also, the money has to be in the bank up front, $5,000 U.S."

"Five grand?" Other prices I'd been quoted were consider-ably more, and those rates were typical. "Are you sure?"

"That's what he said."

"Does his lab test ancient DNA? Older than fifty years?"

"I'm not sure. All I know is, it's new and it's a DNA lab."

"If it's *only* a forensic lab, Ted, then they can't handle the Starchild in the way it needs to be handled."

"I told him everything you told me, and he thinks he can do us some good. I think you should hear him out."

Who was I to argue with that? "I'll call Ray and Melanie."

"What about Mark Bean? Isn't he your partner in this?"

"He just runs the website now. No hard feelings or any-thing, but with him living in Las Vegas and me in New Or-leans, we're just too far apart to work together regularly."

"Okay, then, let's try to meet soon; in the next week or two if you can arrange it with Ray and Melanie."

We were at the end of September.

"I'll push it as hard as I can."

CHAPTER NINE

VANCOUVER

Science in its ideology sees itself as doing a fearless exploration
of the unknown. Most of the time it is a fearful exploration of the
almost known.
                                                    —*Rupert Sheldrake*

O nce I knew the skulls would be cut for testing, I went
to Medical Modeling in Golden, Colorado. CEO Andy
Christiansen and his staff created for me a stereolithographic copy of the Starchild skull that was as accurate
as current technology allowed. In an amazingly precise process, they created a snow-white reproduction that I then took
to Bone Clones in Los Angeles for a spray-paint makeover. It
came out looking remarkably like the original.

Also at Medical Modeling, they allowed me to compare the
Starchild to a dozen child-aged skulls, from a five-year-old
up to a ten-year-old. I found them all to be somewhat lighter
than the adult skull I carried, but still markedly heavier than
the Starchild skull, which towered over them all due to its
large upper cranium. Seeing it alongside the skulls of real
children, even a few with typical deformities, made it clear
the Starchild was as different from them as from the adult.

As Andy Christiansen said, "That is one strange skull."

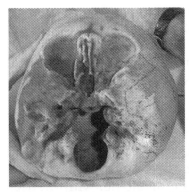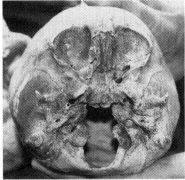

Comparison of stereolithographic copy (left) and real SC skull (right).

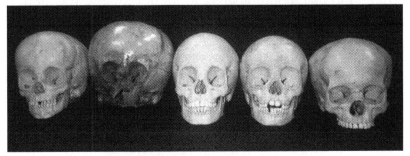

The SC with skulls of children aged five to ten. The two outside skulls have expansion deformities of their parietals, so that area looks similar to SC. The middle one is normal.

Top view of SC with skulls of children. Rear of heads are at top of photo. Outer two have expanded parietals like SC, but with no sign of creases. The middle one is normal.

On October 20, 1999, I joined Ray and Melanie Young, Chad Deetken, Dr. Ted Robinson, and Dr. David Sweet at the BOLD Lab on the campus of the University of British Columbia in Vancouver. It was a full year since Ray and Melanie had been given ownership of the skulls, so by then we were all antsy with the snail's pace of everything involved in the effort. We were excited to finally have this all-important phase of the process—the DNA testing—moving toward a conclusion we now felt would go our way.

Dr. David Sweet turned out to be as charming as his name implied. He was a youngish man, or youngish looking, with dark hair peppered with salt, and boyish enthusiasm to complement an affably confident demeanor. He held doctoral degrees in dentistry and forensic medicine, with a specialty in forensic odontology (dentistry). He was one of four board-certified forensic odontologists in all of Canada. As such, he knew his way around teeth.

Dr. David Sweet examining dental radiographs.

His lab was called BOLD, Bureau of Legal Dentistry, and it was brand new, having been installed as part of a teaching program at the University of British Columbia. Hearing that, I assumed it was why he agreed to get involved with the Starchild. A new lab starting out would make a huge media splash if it could prove any kind of alien connection.

BOLD was one among several other labs (including the Faces Lab in Baton Rouge) whose fascinating work (as mentioned earlier) led, the next year, to the start of the famous *CSI*

TV franchise. BOLD Lab's specialty was "bite mark evidence recovery and analysis, and other aspects of forensic dental science." In addition, it provided "leading-edge research and analysis in genomic DNA cases extending far beyond bite mark evidence, and which involves the use of polymorphic STR loci ensuring optimum results." (Yes, it sounded as impressive to me as I'm sure it does to you.)

The problem was, I had been doing quite a bit of research about DNA labs, and everything I read about them said that the ones for recovery of ancient DNA made the "clean rooms" at computer chip manufacturers look like rubbish heaps. Not a flake of an analyst's skin, or a droplet of breath, could fall in a sample used for testing or it would be contaminated and the result invalidated. It was a highly specialized field that the BOLD Lab did not actually qualify for.

Dr. Sweet didn't say his lab could do such testing. It was equipped for forensic work, not ancient recovery. However, he did think he could determine if the Starchild's bone contained anything worth trying to recover at the higher level. In short, he could provide an intermediate step. It seemed as good a deal as we were likely to get. The price was in our range, and a half-step seemed better than no step at all.

Other than mature red blood cells, every cell in our bodies has a nucleus resting within a glob of cytoplasm surrounded by a membrane. The nucleus also is surrounded by its own membrane. Imagine the cell as a baseball. Strip off the hide. That's the cell wall, the outer membrane. A mile of string is wrapped around a small cork core at the center. The string is the cytoplasm, the cork is the nucleus. Inside the nucleus is the chromosomal package, the mass of DNA given to each individual by combining the DNA of their parents. In humans that package is 46 combined chromosomes, 23 from each parent, containing from 20,000 to 30,000 individual genes—distinct segments of DNA that combine to make chromosomes.

The DNA found in the genes in the nucleus of a cell is called *nuclear* DNA, or nuDNA. In every cell with a nucleus, that nuclear DNA amounts to roughly 3.0 *billion* base pairs, which are four chemical bases (two purines, adenine and guanine; and two pyrimidines, thymine and cytosine) that link in pairs to create the "rungs" in the winding ladder of DNA's famous double helix. That's part one of understanding DNA testing.

Part two is located outside the nucleus, floating freely in the cell's cytoplasm. These are bacteria-sized organelles (a distinct part of the cell) known as *mitochondria*, considered the cellular power plants. (Imagine the blueberries scattered inside a blueberry muffin.) Inside each organelle are "short" segments of mitochondrial DNA, or mtDNA. The mtDNA package contains about 16,000 base pairs, relative to 3.0 billion crammed into the cell's nucleus. Remember that: 16,000 vs. 3,000,000,000—*biiiiig* difference!

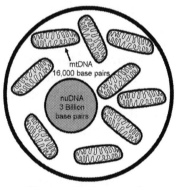

Diagram of a human cell (not shown to proper scale).

Every cell has far more mitochondria than its single nucleus, and mitochondria are durable, more like grains of sand than components of life. Thus, mitochondria are much easier to recover from "ancient" bones than are nuclei. They can even be recovered from fossilized bones, as with fossilized Neanderthal skeletons dating prior to 30,000 years ago. No nuclear DNA had ever been recovered from any fossil, but the geneticists who specialized in recovering ancient DNA were finding nuclear DNA in older and older bones, some much older than the 900 years of the Starchild and its stalwart companion. That race was heating up.

All things considered, it seemed we had a good chance to recover the Starchild's mitochondrial DNA, but the nuclear DNA would be iffy, which presented our most daunting obstacle. All mtDNA was passed from generation to generation exclusively through women, and it passed virtually unchanged. In fact, known mutations were so few that a kind of "clock" could be established based on when each charted mutation entered the gene pool. Utilizing this technique, geneticists in 1987 stunned the scientific world by announcing that humans became genetically distinct around 200,000 years ago. (This has become known as the "Mitochondrial Eve" theory.)

Earlier, we assumed that if the Starchild were an alien-human hybrid, its mother would have to be the human part of the equation. This was based on knowing that a human egg is the most complex part of the egg-sperm equation, so a female is essential as the incubator where the fetus can mature. We further assumed that conception would be artificially created and implanted, the way women like Josi Galante insisted it occurred, rather than through sexual intercourse. Therefore, we would be grateful to have a mitochondrial result to tell us about the Starchild's mother, but what we most needed was nuclear DNA to tell us the genetic heritage of its father.

When animals die, their cells stop functioning and the vast majority decay completely. However, bone cells are durable and don't always decay away to nothing, especially after death in a desert. They can be permanently preserved without decaying or fossilizing. With the skeletons in the mine tunnel in Mexico, "only" 900 years of repose in arid conditions had left them well preserved. The major concern now was whether the Starchild was buried in an acidic soil, which could have led to serious cellular degradation *if*—and it was a highly unlikely "if"—water was present in the soil to saturate the bone and leach any acidic elements into it.

If we could safely assume no saturation for 900 years, the BOLD Lab seemed to have a good chance to recover nuclear DNA, which was its specialty. There would be no detailed analysis, which was the realm of diagnostic labs, but if such a lab knew nuDNA was in the Starchild's bone, they could be sure of sequencing and analyzing both mtDNA and nuDNA.

"The primers we use are for recovering nuclear DNA," Dr. Sweet told us at the meeting. "Think of them as a key in search of a lock to fit. Each one is in the range of 100 to 300 base pairs long, enough to determine whether viable DNA is present. If it is, a full diagnostic analysis is indicated, using primers over 500 base pairs long. At those longer lengths they provide more data about what they recover. So, as I told Ted and then Lloyd, we can provide only an intermediate, but possibly a quite important, step."

Melanie tentatively raised her hand to interrupt. "Can you give us any idea of what we can hope for? Best case, worst case—like that?"

Dr. Sweet complied behind a reassuring smile. "The best case would be a clear pickup of nuclear DNA by the primers in both skulls. A worst case would be nothing recovered from either one. But if they really were found in a mine tunnel, we should be able to recover something from both. That's why I told Ted I thought I could help you. If the provenance story is true, there shouldn't be much degradation."

He then explained loci peaks and other technical terms that would be part of the process, politely answered more questions from Melanie and the rest of us, and finished with a tour of the lab and introductions to his staff. He explained that if we let him and his crew get to work on it right away, we should have an answer in about three weeks, early to mid-November. It sounded like the final mile of a marathon.

Because Dr. Sweet was a forensic dentist, his preferred method of testing with cadavers was by using their teeth. Teeth were some of the hardest parts of a human body, and the pulp from inside intact teeth was often the last part of a corpse to decay. In addition, contamination from outside was virtually impossible, so we all realized that removing tooth pulp would be the ideal way to go about this.

In the case of the adult skull, its top teeth were intact save for the rearmost molar on its right, which was hardly noticeable, so it could easily spare another molar. However, I was reluctant to remove even an obscured molar because its "smile" was so amazing. Though stained like the rest of the bone, the teeth were perfectly aligned, as if by braces, and had not one cavity. (The crowns of each molar were worn flat, no doubt from chewing stone-ground, grit-laced food.)

I found those teeth fascinating.

While the adult skull could lose a tooth and not miss it terribly, the Starchild's detached piece of maxilla had only two exposed, and Dr. Sweet and his staff understood the need to preserve them because specialists could glean an astonishing range of information from them.

Entire species of so-called *prehumans* were named and categorized by slight variations in cusp patterns on fossil teeth, so if the Starchild ever received a full-scale scientific study (the Otzi-like blitz discussed in Chapter Two), I'd be forever condemned for sacrificing either tooth. Because it would be such a loss, I intended to preserve it for as long as I could.

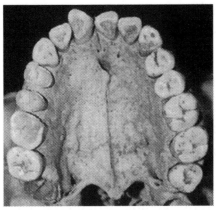

Teeth and maxilla of SFWS. All of its teeth are heavily worn, and its rear molars are flat, due to eating a grit-filled diet that comes from grinding corn on a stone. However, their occlusions (their positions relative to each other) are close to perfect.

[Note: Dr. Sweet disagreed with Linda Faircloth's judgment about the Starchild's age, insisting it be referred to as a "child" because to him it couldn't possibly be an adult. As we've seen earlier, conflicting opinions between experts are quite common, so neither one should be taken as gospel.]

With tooth issues in mind, we agreed to remove the occipital condyles from each skull. They're two knobs of bone in front of the foramen magnum (the hole where a spine enters a skull). It's an area of thick bone that would supply ample testing material, while producing minimal visual impact when viewing the skulls. Besides, the company that painted the skull copy to look like the original, Bone Clones in Los Angeles, had assured me they could replace any missing parts to look very close to the original.

We left Vancouver feeling good, feeling right, on a Saturday night.

During 1999, I managed to get the Starchild skull on several local cable TV shows, the kind few people watch, but I also got it on a popular New Orleans tradition—Frank Davis' *Naturally N'Awlins*—on CBS's local station affiliate, Channel 4. In his usual three minutes, Frank brought out the skull's salient features while also poking his trademark bit of fun at it. Several thousand saw that brief segment and many called or wrote to me (and Frank) about it, but none offered any significant help with funding or testing.

In Tampa, Florida, I was invited to be on Fox TV's *Kathy Fountain Show*, a widely watched local news segment that provided a twenty-minute interview. It created something of a local sensation, so they invited me back not long after. Many more thousands saw those shows, and some of them contacted me about it, but as with *Naturally N'Awlins*, no serious help or support was offered by anyone.

By early October despair was setting in. Then, out of nowhere, a bolt from the blue! When I had barely returned from securing the new stereolithographic copy of the skull, the popular mainstream TV tabloid show *Extra* called to ask if I would agree to be in one of their five-minute segments about the Starchild. *At last!!!* Even at only five minutes, this would be a significant exposure to the entire nation. That *had* to attract the attention of at least *some* scientists and wealthy contributors. I believed this was what I'd been waiting for.

I told the *Extra* producer I'd be leaving for Vancouver in five days, and if we didn't film before then, both skulls would be there and I'd have only the copy I just had made. He put my segment on the fast track and in four days we filmed at a lawyer friend's offices, which gave a much better background for the video than my cubbyhole apartment.

The usual practice was to hire a local crew to shoot the segment, filming my responses to canned questions and shooting B-rolls (where a subject walks or does something in silence so the narrator can speak over them on film). That went off without a hitch, and the four local men who filmed it assured me they had never seen anything as interesting as the Starchild. They promised they'd do whatever they could to convince the *Extra* producers that it was a legitimate matter for scientific inquiry.

I got busy alerting everyone I could think of so they'd be sure to see the show and, hopefully, recommend it to friends and family and—most critically—any scientists or deep pockets of their acquaintance. The more I beat the drums for it, the more I came to feel it was a make-or-break event. I had poured everything—time, money, effort—into this challenge, and now I was about to make it pay off. My despair lifted and I became confident again. I was elated.

It played first on November 12, and I braced myself for a torrent of phone calls (I was listed in the directory to be easy to contact) and emails from around the country. However, as with smaller shows in smaller venues, nothing happened. Dozens of extra emails, as usual, congratulatory or offering curiosity about some aspect or other of the skull—but no offers of the serious help we needed to forge ahead.

A few weeks later, as 1999 closed out and 2000 loomed large, *Extra* announced it would air a millennium-ending one-hour special to highlight their best shows in various categories for the decade of the 1990s. I was called again by the same producer I'd been dealing with, who told me the Starchild's segment won the UFO category hands down, and it would air again as a repeat in the last week of the year, decade, and millennium. This was great news, of course, a very welcome second chance—but I had reservations.

"What I just don't understand," I said to Ralph, the producer, "is this: if the Starchild segment really was so great, why didn't anybody react to it?"

"*Not react?!*" he exclaimed. "We had calls and emails up the wazoo! That's why we're replaying it. It was fabulous TV!"

"Then why didn't anybody call *me*? Why no *help*?"

A pause hung on his end. "Listen, we're friends now, right?

I can be straight with you. So let me say that when I first heard about you, and was assigned to arrange filming you, I assumed you were just another wacko. But seeing you talking in the raw footage, then editing your piece, I've come to respect you and what you're trying to do with the Starchild."

"Well, thanks, I appreciate that."

"The problem is, just about everyone here thinks it has to be some kind of fraud, or a clever fake. Everyone makes the automatic assumption that if it was a real human-alien hybrid, you wouldn't have it. Nobody would be allowed to hold onto something so explosive. It would be confiscated."

That was surprising. I thought that bit of "knowledge" was reserved only for those on the inside of UFOlogy. I hadn't realized it was so widespread. "They can't be any surer about it than I am. They're probably assuming it's a fake, too."

"It's their job to stay on top of all things alien. If they thought for a moment it was real, they'd take it from you."

I recalled my gamble of announcing it at Laughlin. That was turning out to be a good move. "Maybe they're waiting to see what I can come up with."

"To see if you can prove it? That's a bit risky, don't you think? I mean, if you *do* prove it, the horse is out of the barn. There's no way they could ever put it back in."

"You're right, they'd just have to live with it."

Another pause on his end. "Listen, Lloyd, I believe you, and I think you believe it's the real deal. But it has a serious perception problem. Actually, *you* have a serious perception problem, and I don't know how to help you get around it."

"Well, I appreciate your being honest . . . I think."

He chuckled softly. "Hang tough, my friend. It looks like you have a long, rocky road ahead of you."

At times like that the road seemed long, rocky, and, worse than either of those, heading toward a dead end.

Long before *Extra* called, I was running on empty. I'd gone flat-out for six months, the length of time I told Ray and Melanie Young that I would need to secure interest from mainstream scientists. Nothing I predicted had panned out. Radio and email appeals for contributions for testing produced only meager support. Money became a central issue.

Researchers in the alternative knowledge community are prevented from accessing funds from any mainstream source because the grant system is structured to exclude nontraditional inquiries or testing. This means the only source of funding is ordinary people with an interest in the work, which creates an extremely small pool of money for all researchers to draw from. Competition is intense because each person has only so much to give to any one cause. While we did receive many small contributions that were helpful and appreciated, in the larger scheme of what we were up against and what was needed, it simply was never enough.

There was also an emotional aspect to what the Starchild might mean. I slowly came to realize that the choir we preached to in alternative circles was heavily invested in keeping our own gospel alive. For them it was a passion to which they dedicate time, energy, effort, and sometimes money to causes that touched their hearts. However, the primary satisfaction for the majority seemed to lie in the ceaseless struggle to have their views heard and accepted by the mainstream, which stubbornly resisted any accommodation.

That being the case, it was a gigantic emotional step for such people to contemplate having all doubts regarding alien existence removed by the Starchild. Down deep inside, most didn't really want that doubt eliminated. For them the fun was in *playing* the game, not winning it. Everyone loves a good mystery, but after it's solved, nobody cares.

The *Extra* repeat was heavily hyped as a special, and the Starchild skull was shown in many advertising blurbs. I felt certain it would finally be the big break we needed. Surely its status as the top UFO story of the decade would get the attention of scientists all over the country, if not the world, and offers of help would pour in. I started breathing a bit easier.

Despite all the hype and anticipation, no calls came in from scientists, doctors, financial backers—nobody. Only another increase in emails, like the one it brought in November.

We threw a party and nobody came.

Meanwhile, as all that went to hell in a hand basket, on November 2 an email came from Dr. Sweet with an interim report on the testing at the BOLD Lab. He explained that the condylar samples were removed as planned and extensively cleaned, first by abrasion of their outer surfaces, followed by rinsing and soaking in various sterilizing bleaches.

He further stated: "Human genomic DNA was extracted and typed from the adult skull using the screening test Amp-FISTR-Blue (at 3 genetic loci), plus amelogenin (gender determination). Results reveal that the DNA extracted from the adult skull is from a female person. A genetic profile at three forensically significant loci has been obtained."

Sounded like her DNA was easy to recover. *A good sign!*

"Human genomic DNA was extracted and typed from the child skull using the same screening test. Unfortunately, the DNA profile is a mixture of at least three people. The amelogenin locus (gender typing) determined at least one DNA contributor was male (the result is a mixture of a male and a female). This indicates contamination of the specimen by DNA originating from other people. Despite our best efforts to decontaminate the surface of the specimen prior to DNA extraction, multiple profiles were obtained."

This was a disheartening, though not surprising, result. During the meeting with Dr. Sweet, he stressed that contamination was the #1 gremlin in DNA analysis, and every lab worked diligently to minimize it. With ancient DNA recovery, cleanliness had to be taken to ridiculous extremes.

Because BOLD wasn't equipped to maintain the antiseptic levels of a true ancient DNA lab, it was probable the contamination came from one or more of its employees and/or student trainees from the University of British Columbia. However, many other human hands had held both skulls at one time or another, so some of them could just as easily have left a residue that, for some reason, wasn't removed by the extensive cleaning procedures.

"Based on my experience with trace evidence samples from calcified human tissues," Dr. Sweet wrote to me, "I recommend additional DNA tests be undertaken." I was glad to grant him permission to have another go at it, but now, after the first-round failure with the Starchild, he felt it was time to sacrifice one of its two teeth to be as certain as possible of securing a verifiable result.

I had deep reservations because of what losing a tooth meant, and because if contamination had occurred once, it could just as easily occur again. However, he was adamant that this was the best way to go, so I really had no choice. I'd stepped onto another of life's treadmills that quickly became impossible to slow down. I agreed to let him remove the back-side premolar to sequence it, and in several days he replied.

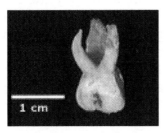

Rear molar extracted for DNA testing (destroyed). Roots are stout and strong, not at all like normal roots of baby teeth, which are reabsorbed before the tooth falls out.

"Unfortunately, the results of the PCR amplification of the sample recovered from the baby tooth taken from the piece of right maxilla (upper jaw) was negative. No profile was obtained. This may be due to factors in the environment (post-demise), including humidity, acidic soil, UV light, heat, etc. I think it will be valuable to harvest another small sample from elsewhere on the exhibit and attempt another amplification. This is the only way we will be able to obtain any result."

Nothing??? How could *nothing* come from inside a *tooth?* If one part of the skull was a kill shot for DNA extraction, that was *it!* If they couldn't recover from *inside a tooth,* the odds were that no DNA would ever be recovered—*ever!* This meant we'd have to try to establish the Starchild's parentage through other means, through logic and reasoning and fair debate, things mainstream science would never allow.

In short, for our side it was about as bad as news could get. It was downright catastrophic.

Devastated, with nothing to lose and not much to gain, either, I agreed to let them remove the right mastoid, a more visible knob of bone bulging down behind human ears, which would provide a thick testing sample. While I was on the phone arranging this with one of Dr. Sweet's assistants, he mentioned that the Starchild's bone was proving unusually hard to cut, which entered my brain but didn't lodge as it should have. Later it would be meaningful, but the first time I heard about the skull's unusual hardness, I wasn't giving it my fullest attention because I could feel the whole project slipping away with the inability to recover any DNA.

On November 12, the same day the *Extra* segment played for the first time on national TV, I received this report from Dr. Sweet regarding his analysis of the mastoid:

"After the customary number of PCR cycles (28), there was a weak gender profile from the second bone sample taken from the child's skull (mastoid process). Other alleles had 'dropped off,' which is usually a result of degradation of the genomic DNA. The approach when this occurs involves reamplifying the sample with 5 additional cycles in an attempt to produce a result. Once these additional 5 cycles were performed, the outer limit of current amplification technology, only one locus showed a profile: amelogenin. The result is X-Y, which tells us two significant things: first, the child was male; second, the DNA is human. Unfortunately, because we do not have a profile from other loci it is not possible to conduct a paternity test against the genotype of the adult skull to determine if they are biologically related."

By then I had met with enough experts, psychics, and UFO buffs to know they all were tossing mental darts at the skull, hoping to hit somewhere near it. That seemed to be everyone's modus. So, by the time I went to Vancouver, I was close to certain the Starchild was a human-alien hybrid—maybe 90 percent hybrid, 10 percent it's a ten-billion-to-one deformity. I had swung 180 degrees from my original doubt to a certainty that I was correct, so I fully expected *any* DNA results to validate my opinion. Therefore, from my standpoint, this result was well beyond catastrophic.

It was the very last thing I ever imagined I would hear.

While I couldn't reject the BOLD report outright, I could—
and very much did—doubt it. With contamination on the first
test followed by failure to recover anything from their supposed specialty, a tooth—*a tooth!*—how could anyone be confident in any result they produced? I certainly wasn't, and
even though this result was now a part of the Starchild's official record, I didn't believe it should be. I contended that the
human DNA Dr. Sweet felt he found in the skull's bone didn't
automatically rule out the possibility of it being a human-
alien hybrid. Let me explain why.

Any human has 22 individual chromosomes, termed *autosomes*, and 1 sexual chromosome (an X or a Y) contributed by
each parent. That's 23 from each parent, giving a total chromosome count of 46, with males having an X-Y combination
and females an X-X. Females are the default member of our
species, so at the moment of their conception there will be a
random shutdown of one of their two X chromosomes, and
they will live utilizing only one of the two. But every female
will continue to carry the other one, and it can be expressed
by any of her progeny that inherits it. Males, on the other
hand, have to utilize both of their sex chromosomes (X-Y) because they need the X to construct the default form of their
body plan, and then a Y modifies it to turn them into males.

With that said, let's consider a female human-alien hybrid.
At conception, one of her two X chromosomes shuts down (it
could be either one, the process is random) and will stay inactive
for the entirety of her life. So if she's a hybrid from an alien
father and a human mother, and if the X she inherited from
him is the one that shuts down, she can live a normal life
that can include reproduction because, statistically, half her
eggs would carry the alien X gene, which wouldn't reproduce
with humans naturally, while the other half would carry her
mother's normal human X.

This isn't the case with male offspring because, as noted
above, males require both the X and the Y to be present and
functioning before they can mature and reproduce. Neither
X nor Y is, or can be, shut down. If a male is born and lives,
it is because his X and Y chromosomes successfully worked

together during gestation to fashion a male's anatomical morphology out of the default female morphology. Without such a perfectly functioning symbiotic relationship between the X and Y chromosomes, a male could not gestate to birth.

None of the other 22 autosomes in either sex are so closely intertwined in their functioning, nor are the two X's of a female. The human Y's responsibilities for forming males are so highly refined that any malfunction is unlikely to produce viable offspring. However, if a Y from an alien being were to be matched with an X from a normal human, and the alien being was genetically quite similar to a human, then a viable offspring might be produced but it would be sterile, like a mule—viable but sterile.

Any reputable geneticist like David Sweet would have to base the analysis of his results on those well-established truths. However, he would also be required to assume the Starchild was conceived in a "normal" sexual union (which seemed improbable but was not impossible). If that assumption was correct, it meant both the X *and* the Y chromosome in the Starchild's cells had to come from normal humans, which meant it *had* to be a normal human, too.

Dr. Sweet did not—and professionally he could not—consider that the Starchild was the result of an in vitro sexual union (similar to how we do it now in a petrie dish). Yet anatomical and biological necessity might make such a process imperative if the Starchild was, indeed, the hybrid offspring of a human mother and an alien father. Its hybridization may well have resulted from an in vitro conception, followed by a timed implantation of the fertilized egg into a mother's body.

For as unlikely as it might sound, orchestrating conception between an alien and a human could be done by injecting a human male's sperm with a functioning Y chromosome into a human egg with its X chromosome still in place but the other 22 autosomes removed and replaced with alien ones. Choosing to create either a male or a female gets even more complicated and does not need to be dealt with here. The point is that such hybridization *can* be done, especially by anyone who knows what they're doing in a genetics lab.

Despite Dr. Sweet's misgivings, the Starchild's sex chromosomes (X-Y) may have been no different from the typical human pattern because both, by planned intention, could have been taken from "lab-rat" parents. Likewise, some, most, or all of its other 44 autosomes could have been altered by genetic manipulation to create its astounding array of physiological differences. (These, too, are issues that can be resolved by an all-out, Otzi-like campaign of discovery.)

For all of these reasons, I could not accept that the Starchild *had* to be completely human simply because it was a male that responded to a normal male's amelogenin profile. I expressed those reservations to Dr. Sweet, who sympathized with my profound disappointment that the result hadn't been what I anticipated. He offered to do a final round of testing on a sample taken from the parietal bone (the upper right rear of the skull), an area that could never have been touched by fingers on its inside surface, and whose outer surface was covered by a protective coating of shellac.

All along we had hoped to avoid cutting into the dome of the skull in order to keep its unique outline as intact as possible, but now we were down to a very taut wire and had to pull out all the stops. I reluctantly gave permission to run test #4 on the piece of parietal, not knowing what could be done to halt this careen toward final disaster.

On December 2, a month since the BOLD Lab's first report, Dr. Sweet sent me his analysis of the parietal sample.

"After receiving permission to proceed, a 4 cm by 4 cm section of parietal bone was harvested from the lateral aspect of the right side of the skull. Rigorous decontamination procedures were used to eliminate any contaminating DNA. Subsequently, 5.5 grams of powder were produced by cryogenic grinding. The sample was divided in two, and DNA extracted and purified from one-half the total amount. It was determined there were 200 picograms of DNA present in this relatively large sample. (Ideal amount of target DNA for PCR is 1,000 picograms.) Thus, again, it appears there has been considerable degradation of the DNA over time due to environmental conditions at the site of discovery or during storage and/or transportation.

"PCR-based amplification of this DNA trace produced an X-Y (male) profile, but did not result in any supra-threshold results at other significant loci. (No peaks.) The second sample was then processed and concentrated. PCR-based amplification of the DNA produced the same result as the first half. That is, X-Y (male) and no peaks at other genetic loci.

"Due to the strict cleaning regimen employed with this sample, it is my opinion the DNA that was isolated and tested was not from exogenous, contaminating DNA. The result appears to be due to the age of the skull; the genomic DNA is too degraded to provide a complete profile. The sex of the decedent is verified as male. DNA traces recovered in each of the numerous tests performed in this laboratory responded to the human-specific probes.

"The following question arises: Can DNA be used to evaluate, assess, or diagnose the etiology (cause) of the unusual shape and appearance of the child's skull? Unfortunately, this lab deals only with STRI loci with forensic significance—they are nondiagnostic loci. The specimen would have to be tested by a laboratory that focuses on diagnostic genetic loci if you were to consider attempting to identify a potential genetic cause for the unusual appearance. Further, it is predicted that the diagnostic facility will also find the same difficulty isolating and extracting sufficient DNA."

In short, those who felt the Starchild represented something significantly different from human life were screwed. Not in the sense that Dr. Sweet or his lab had done anything out of line. I'm sure they did their best, given the circumstances. However, the results they provided could not have been more detrimental to the idea that the Starchild might be a human-alien hybrid. In fact, they had eliminated that possibility. In their view, if it was a male, it was human. There was no ambiguity, no wiggle room. Done. Over.

Personally, though, I simply could not accept it. By then I had been through too much and learned too much. Something *must* have gone wrong . . . *but what?* I had no idea and didn't know where to look. All I knew was what I *knew.*

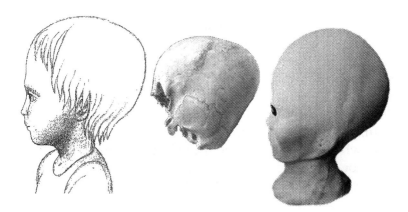

Comparison of SC's cranial profile (center) with a human-alien hybrid child (left), and a Grey alien (right). This obvious similarity was one of many factors that contributed to Lloyd Pye seriously considering the possibility that SC could be a human-alien hybrid.

The Starchild skull was wildly different from ordinary human skulls. Dr. Ted Robinson had found a study (see his report in Appendix I) that showed if 30 points of reference were measured on 100 typical human skulls (i.e., from the brow ridge's outside corner to the ear canal opening, or from bridge of nose to point of chin), those 100 combined measurements would provide a statistical tabulation that could establish a reliable mean average for all 30 points of reference.

This large group of measurements produced a bell curve distribution of normal elevation and range. Thus, 3 standard deviations from the mean of this data set would contain roughly 70 percent of all measurements. Also, 5 standard deviations would contain 99.9 percent of all measurements. Anything beyond those 5 standard deviations would conclusively move beyond variation into the realm of deformity.

Naturally, among individuals there could be differences. A Watusi's 30 measurements might sharply contrast with an Eskimo's, but neither Watusi nor Eskimo would, on average, lie more than 2 or 3 standard deviations (SD) from the mean. Extraordinary would be 4 SD away, while 5 SD would suggest deformity rather than acceptable variation.

Because the lower half of the Starchild's face is missing, many measurements along its 30 points of reference could not be taken. However, those that could be taken (corner of brow ridge to ear opening, width across parietals, crown to inion, etc.) were consistently beyond 5, and often in the range of 10, standard deviations from the established norms. This positions the Starchild skull at the farthest edge of deformity, where nothing remotely human dwells.

That being the case, I felt it was reasonable to believe the Starchild skull was unlike anything ever seen before by anyone anywhere. It was just too incredibly weird in far too many ways to be entirely human. That was something I seemed to *feel* more than I knew, but I felt it powerfully.

Despite my strong misgivings about the BOLD Lab's report, I didn't have a public leg to stand on. I had announced on several radio shows that the DNA testing was underway in Vancouver, donations had been solicited to defray testing costs, and contributors and millions of others were waiting to hear the results without regard to their impact on me or on the Starchild Project itself. Everyone was ready to reach the end of this already stretched-thin ordeal.

I bit the bullet and reported Dr. Sweet's results to the loyal following on my extensive mailing list, then took to the airwaves with all the top radio hosts—Art Bell, Jeff Rense, Hilly Rose, Laura Lee—to discuss it. However, each time I did, I was careful to stress that all four analyses were done in a forensic dental lab not equipped to use the most stringent decontamination techniques. Stringent for a forensic lab, but not for a properly accredited ancient DNA lab.

I was also careful to note the amount recovered was only one-fifth of the minimum usually required for forensic investigations. This sample, 200 picograms, was a dangerously small amount on which to hinge a potentially momentous result. And, try as I might, I could not get across in a cogent manner the idea that an in vitro conception could leave an X and a Y sex chromosome visible to DNA primers, yet the other 22 autosomes could be alien. In radio terms, crafting a scientific argument around such complex material was a guaranteed "snoozer." (I've been warned it is likely to be the same in this book, though I hope that's negative thinking.)

In any event, no matter what I said to ameliorate the BOLD Lab's result (what some might call "spinning" it), nobody heard me. What everyone heard (I now realize) was what they all expected or wanted to hear: *The Starchild is human.*

Both skulls were with Dr. Ted Robinson for much of 2000 while he coaxed several medical colleagues to examine the Starchild and render official opinions about it. [These are in Appendix I.] At the end of Ted's work with the skulls, both were returned to Ray and Melanie Young in El Paso, where they remain under lock and key. I kept the stereolithographic copy and the piece of maxilla, the latter to show, whenever it became necessary, that the skull's bone was indeed genuine.

As the millennium turned, so did the Starchild's fortunes—all for the worse. My hopes were momentarily raised when a call came from *Extra* saying the Starchild segment would be replayed as the best UFO piece of the 1990s. However, the lack of response—which might be attributable to my numerous radio appearances—sealed the deal. The Starchild Project was on life support, and it was time to pull the plug.

Without consciously realizing it, I had followed Rudyard Kipling's stern advice in "If," his beloved poem about what it takes to be a man. I had made a heap of all my winnings, such as they were, and risked them on one metaphorical turn of "pitch and toss," a game of chance mentioned in the poem. If you lose, Kipling advises, *start again at your beginnings, and never breathe a word about your loss.*

In that sense I have to admit I wasn't much of a man, because I did breathe quite a few words about my blindsided loss, perhaps the least offensive of which was Mark Bean's colorful exclamation, "Holy Maloley!"

Then again, Kipling goes on to say:

If you can force your heart and nerve and sinew
To serve your turn long after they are gone,
And so hold on when there is nothing in you
Except the will which says to them: Hold on!

That much I did . . . barely.

CHAPTER TEN

SECOND WIND

It is the business of the future to be dangerous. The major advances in civilization are processes that all but wreck the societies in which they occur.

—*Alfred North Whitehead*

They say time heals all wounds, but it wasn't doing me much good after those inflicted by the BOLD Lab, the *Extra* fiasco, and the consistent disinterest of celebrities. I slowly, inexorably sank into depression as every roll of the dice was snake eyes. Making matters worse was my plummeting freefall into debt, making it increasingly harder to make ends meet. Finally, back squarely against the wall, I turned to an old friend from college, an ex-teammate on the Tulane football team with whom I'd recently reconnected.

Pat Snuffer, owner/operator of all *Snuffer's* restaurants in the Dallas-Ft. Worth area, came to my rescue, not out of any belief in the Starchild, or in my promotion of Intervention Theory (a primary thrust of my work prior to the Starchild entering my life), or hominoids, or anything of that nature. It didn't interest him. He was simply willing to help a friend in need because he'd seen me do a presentation when things were going my way, and he knew what I was capable of when I was on top of my game. Pat was the difference in my getting through 2000, 2001, and 2002. It wasn't that I didn't do anything for myself. I traveled, lectured, sold books, and did radio interviews. Much of that carried on as before. However, invitations to speak at the big conferences stopped coming. I was damaged goods, the guy who didn't make the Starchild happen. I was old news now, operating at anywhere from one-fourth to one-third capacity, depending on the month.

If you should ever find yourself in the Dallas-Ft. Worth area, please do us both a favor and eat at one of Pat Snuffer's restaurants (they're all called *Snuffer's*). He's owed a ton of good karma, and his places make superb hamburgers.

As the snake eyes I kept rolling sunk my financial fortunes deeper into the abyss of debt, a new girlfriend entered my world, emotional life-preserver in hand. Karena Bryan was a dark-haired, dark-eyed beauty in her mid-thirties, with an ebullient, brassy personality and certified credentials as one of the world's few female shamans.

Karena Bryan

Karena lived and worked in San Francisco, while I stayed rooted in New Orleans, requiring a clearly defined "long-distance" relationship. Luckily, she had the resources to ensure that we could spend a week or two together every six weeks or so, which left us dealing with the many strains of intermittent separations, though never to the breaking point.

An unexpected windfall came in August 2001, when I was contacted by Beyond Productions, a topnotch TV production company from Australia. By then Globo TV, a large international news distributor, had taken the *Extra* segment worldwide, so when the Australians saw it they felt it might support an hour-long documentary. They asked if I would cooperate if they got a go-ahead from The Learning Channel (TLC). It was all I could do to keep from bouncing off my apartment's low ceiling as I jumped around. *An hour! A whole freakin' hour!*

Two weeks later they called again to tell me the "suits" running TLC were uneasy about a full hour on such a controversial subject. The suits agreed to finance the show only if the Starchild skull was just one part of an in-depth look at unusual skulls from around the world. So the Australian producer, whose name was Brian, asked me to tell him about other available strange skulls they could focus on.

"Other than the Starchild, the only possible nonhuman ones I know about are the coneheads in Peru. Museums there have hundreds of them on display and in storage."

"Those silly aliens from France? They're not real!"

"The TV coneheads aren't, but they were modeled on those in Peru, which are real, true, bone skulls like the Starchild, but exactly the opposite. Their facial structure resembles humans, but the bones are much thicker and heavier than usual, and the brain volumes are twice as large as ours and held in a cone-shaped cranium. They're as weird as the Starchild."

"Exactly what we need! Now, what else is out there?"

"Nothing, Brian. Those two kinds are all I'm aware of."

Silence on his end, then a heavy sigh. "Listen, mate, we can't fill an hour with only those coneheads and the Starchild. You have to help me out here."

"You could go to the Mutter Museum in Philadelphia. They specialize in bizarre birth defects, although that's not what the Starchild and coneheads are about. Those two types could represent something not entirely human."

"That's why we hoped to sell the Starchild to TLC. Unfortunately. . . ." He sighed, letting it hang. "They don't want anything *too* sensational . . . just sensational enough."

"For what?"

"Viewer satisfaction. Our goal is to titillate, not frighten."

Huh?

"Now, Lloyd, there's one more thing," Brian went on. "This is always a dicey subject with people like you, but I have to say it. We need someone with academic credentials who will go on air and give us an opposite view."

I had danced this dance a few times already. "You mean they'll tell the audience that everything I've said is wrong and it's nothing to be concerned about."

"If you want to put it like that, yeah, we need a counterpoint . . . a different take on your material."

I knew that all TV shows about controversial topics such as UFOs, aliens, ghosts, hominoids, etc., always had one or more scientists on camera heavily pushing the mainstream party line. But I never understood why it was necessary.

"Then what's the point of going to all the trouble to make a show out of it? I mean, if a naysayer can come in at the end and negate my whole message, why bother?"

"It's not necessarily the best publicity for you," he admitted, "but like bad breath, it's better than none at all. I suggest you go along with how we have to do it. And, remember, it's the same everywhere, across the board."

"Then I repeat my question: What's the point?"

"To *entertain!*" he growled. "That's what we do. We produce shows that contain great material for teasers, the few-seconds-long adverts you see in the run-ups to airings. Good teasers snag audiences like nothing else. But once we have people watching, our job is to entertain them, not send them off to bed scared half to death that UFOs or aliens or ghosts are real, or that something like your Starchild skull might actually *be* a human-alien hybrid."

"But that's the truth! It might be!"

"You can say that on air, I promise. But I also have to tell you that we can't run a show or a segment like this without a disclaimer given by someone who'll put everyone's mind at ease so they can go to sleep that night and wake up to the world as they've always known it. That's our job."

"Your job sucks," I groused.

"So does a few hundred irate calls and emails to the producer, to the channel, to everyone with a hand in it. We've all been down that road and learned our lessons. Now, are you with us on this, Lloyd, or do I give it a pass?"

(What would *you* have done?)

"Call Dr. David Sweet, in Vancouver, Canada. He'll tell you it's a deformed human."

I could almost hear Brian gloating all the way from Australia. "Right-oh, mate! Then we're on for filming in late October, airing June or July next year."

"Yeah . . . fine . . . whatever. . . ."

Mystery of the Skulls aired on July 9, 2002. It opened with a focus on the coneheads of Peru, segued into an analysis of the Starchild, then moved into a study of living people with unusual skulls. No mystery, just deformity, but better TV than the Mutter Museum would have provided.

For my part in it, I was interviewed in a dim room and lit from below, making me look like a wild-eyed zombie. Even my mother thought I looked disreputable. Dr. Ted Robinson was also interviewed to give a positive spin on what the skull might be, but he was filmed in a brightly lit, extreme close-up that made him look more wild-eyed than I did.

Only Dr. David Sweet was shot in normal light at a normal distance, making him look and sound like the most rational person in the segment as he said what he believed: that despite how exceptional the Starchild seemed at first glance (he admitted he initially felt it could well be alien), the DNA results he obtained were clear and unambiguous.

It was a human male—period. He left no room for doubt, and because he was the show's skeptic, he had the last word.

As with the *Extra* segment, we threw a party and nobody came. No huge boost in emails, only a few extra phone calls, and most critically, no offers of help from any quarter. I'd foolishly placed my hopes on this show because I had nothing else to count on, nowhere else to turn for support. If this failed, I'd be the proverbial blowed-up peckerwood, which was coming to pass. *Sleep tight, everyone.*

With *Mystery of the Skulls* leaving an indelible impression on a national audience (and later, on a worldwide audience) that the Starchild skull was merely an oddity rather than a profound mystery, the path for me to find out the truth about it was effectively compromised. Whenever anyone asked me about it on radio interviews, I'd invariably mention the vast room for doubt in the BOLD Lab's conclusions. Basically, though, nobody wanted to hear from me about it. It was old, stale news. Finished. *Kaput.*

To add insult to injury, by then most regulars on the alternative knowledge circuit had seen my slide presentation based on my book, *Everything You Know Is Wrong*, and while it was considered one of the best available, its expiration date loomed. If I didn't get on a new path soon, I'd be marginalized, then trivialized. By late September, I was out of options.

"I'm going to declare bankruptcy and start out fresh in 2003," I told Karena, who was visiting me in New Orleans. "I have to stop this downward spiral and get my life back on track. I need to move in another direction."

This was no surprise to her. She knew I was working myself to a frazzle and worrying myself sick just trying to pay the interest on $50,000 of credit card debt. The principal wasn't going down a bit, and I hadn't made ends meet without Pat Snuffer's help in well over a year. It was time to face up to this abject failure and deal with it.

"You mean quit everything?" Karena asked. "All of it? The Starchild? Your book? All you've been doing?"

I nodded. "I want to wrap it all up by Christmas and start writing the magnum opus right after New Year's Day."

The magnum opus was something I'd started thinking about earlier that year. I'd been told by several conference promoters that I needed to publish a new book. The prolific outputs of colleagues like Alan Alford, Laurence Gardner, and Graham Hancock put me to shame, so I settled on the idea of doing a new "big" book of the kind those men produced. After attending conferences and listening to many other researchers present their information, I knew I could create a new volume that would greatly enrich what I'd put into *EYKIW*. The opus could genuinely become magnum.

"If you're going bankrupt," Karena went on, "how do you plan to pay for the time to write it? You'll need at least a year to do it, don't you think?"

"At least," I agreed, "but it's the kind of thing I can shop to investors, people with money willing to try to make more money by betting on me."

She had run her own business for years, so she wasn't without that kind of acumen. "Hon, I don't see much of a market for investing in a writing proposal."

"Actually, it's a pretty good risk. I'm sure I can turn it into a viable business deal."

With the Starchild I couldn't do that. I had to beg for money like a bum, free and clear without strings, because if I turned it into a situation where investors expected a return for their money, the Project's credibility would be shot. As had happened with the *Alien Autopsy* film, among many other films, relics, and artifacts, the moment any degree of financial interest was openly insinuated into the process, nothing about it could be trusted from that point forward.

"You see," I went on, "I can sell shares in the book's future revenue stream. Let's face it, all over the world millions of people know my name because I've done so much radio and TV and magazine publicity. So what I can offer investors is name recognition and a long track record as a writer. Plus, in four years I've sold nearly 30,000 copies of my last book, self-published. Put me in that light and I *am* a good risk."

"Okay, I can see that," Karena said. "But what about the Starchild? And Ray and Melanie? How do you plan to get out of your commitment to them?"

"I'll tell them like I told you. They know I've given it my best effort, and they know I'm in crisis mode now. It's just time to accept my failure and pass the skulls on to someone else."

"They always say you're the only one they want to do it."

"Things change. I'm changing . . . maybe they will, too."

She considered, then said, "When will you tell them?"

I walked to the phone and picked it up.

In ten minutes it was done. Ray and Melanie protested, but they knew how difficult my life had become since the BOLD results were posted on the website. They knew I couldn't go on. They thanked me for carrying the load as far as I had, and promised to inform me about whatever they did next.

An hour later I posted an announcement to the extensive Starchild mailing list, explaining to everyone that it was over, I was finished. I went to bed that night feeling an odd mixture of regret and relief, embarrassment and pride. I had busted my butt for it and failed, but I had given it all I had. I didn't think anyone else could have done more.

Better, probably, but not more.

Emails flowed in from dozens, then hundreds of people, thanking me for the effort I gave, wishing me well with the new book. I was breathing again, feeling alive again, making a list of people I could approach with the "invest in my new book" idea. (That list didn't include Pat Snuffer. If I did manage to put a deal together, he was already solidly in it.)

Not long after, on a crisply cool day in October, a year since the Australians had filmed my part in *Mystery of the Skulls*, a typical email came among that day's dozens, from a woman in England I had never heard of.

Belinda McKenzie wrote from London to ask why I had not been able to raise enough money to continue. As was my habit, I answered politely that money inevitably became a crux issue for virtually all researchers in alternative knowledge. We seldom made enough to live on from speaking and selling our books, but a single person with a strong message and no spouse or children to support could sometimes manage it, as I did in 1998 and much of 1999. Now the Starchild had driven me under, so I had to abandon it. I assumed that would be enough to suit her.

It wasn't. Belinda responded by asking what tests were necessary if money were no object. I told her our fundamental need was another round of DNA testing in a lab specialized for the recovery of ancient bone. Those costs were far more expensive than ordinary DNA testing, but they could tell us things about the Starchild that nothing else could. We had to establish the genetic backgrounds of both of its parents, which could only be revealed by recovering nuclear DNA.

Belinda wrote back to ask what was needed for that testing and for me to live through getting it done. I provided her those estimates—excluding my current debt load, which the bankruptcy would resolve—and she said she'd check on them and get back to me. In a few weeks she did, telling me she was satisfied I was on the level and asking where to send the money. That set my hands shaking so badly I could hardly type the address of the Starchild Project's bank account, but I got it done, and in a few days everything was back on the original track I had started on in late February of 1999.

By New Year's Day of 2003, when I anticipated starting my magnum opus, the Starchild skull's DNA testing was funded and I was back in the business of finding out what it was.

For three years I hadn't bothered researching or contacting ancient DNA labs because I had no money to pay for their services. Now I found a very different landscape than had existed in 1999. Such labs were now proliferating worldwide, and I soon found a new one, Trace Genetics, just coming online at the University of California at Davis. Two young geneticists, Ripan Malhi and Jason Eshleman, were behind it, and they seemed ideal for what needed to be done with the Starchild—*if* they would agree to work on something so certain to rankle their peers. On the other hand, any new lab needed to cover its startup costs, and I was at long last in a position to apply some financial leverage.

After a series of emails and phone calls, I convinced them I wasn't a crackpot and that the Starchild skull was something legitimately requiring the kind of expertise they could provide. As I expected, their need for clients and money was a motivating factor in their participation, at least initially. However, I believe that when they saw the skull for the first time, they became truly committed to solving its mystery.

I agreed to deliver both skulls to Davis on February 10, 2003, eight days short of four years to the day from when I first laid eyes on them. Since the 9-11 attack 18 months earlier, I could no longer call them "movie props" and freely get them through airport security checks. The last thing I could risk at that point was confiscation by the government, so I drove to El Paso to pick up the skulls from Ray and Melanie, then drove on to the San Francisco Bay Area to pick up Karena at her home. From there we anxiously completed the two-hour trip to U.C. at Davis, emotions surging between delirious anticipation and grinding fear of failure. *Holy Maloley!*

We made our way to the Trace Genetics lab to introduce ourselves to Ripan Malhi and Jason Eshleman, then we went to lunch to get to know each other a bit. Both men were in their early thirties but looked younger. Ripan, whose heritage was Indian, looked about twenty, if that. Jason was a lanky fellow with quick, darting movements that seemed always directed toward a goal, even if nothing more than picking up a fork. Karena and I liked them right away, and we felt our first impressions were reciprocated by both young men.

During lunch I told them what I felt they needed to know about the skulls and what we hoped to learn about them.

Ripan Malhi (left) and Jason Eshleman (right) of Trace Genetics in California.

In turn, they told us what we should expect from what they could do, using present technology. We were impressed to learn they'd worked on the famous Kennewick Man skeleton found in Washington State, which the U.S. government seized to placate local Native Americans intent on protecting their lucrative casino income.

By the end of lunch we were clear with each other. They realized they'd be working on a relic with genuine historic potential, and we knew the steps they would go through to try to secure the answers we needed.

We all returned to the lab to settle into the task at hand.

The first objective was to remove samples of each skull for testing. The adult was missing only its occipital condyles (from beside its foramen magnum), which David Sweet had taken to determine its humanity and sex. The Starchild, however, was missing both occipital condyles, its right mastoid (from behind its right ear), and a two-inch square taken from its upper right parietal (covered over in Vancouver with a plaster patch). The rear molar on its detached piece of maxilla was also gone, but no one would be taking the remaining tooth for any reason, which I made clear.

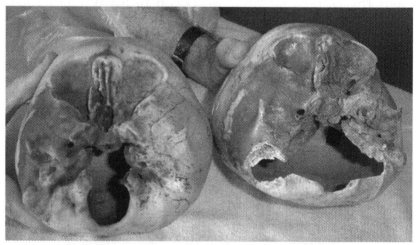

SC copy (left), real SC (right), illustrates the damage from testing at BOLD Lab.)

To my dismay, Jason and Ripan made it equally clear they had no compunction about taking any of the adult's teeth. I wouldn't let David Sweet take one because they were so perfect, but these guys gave me no choice. "Teeth are best for what we do," Jason explained. "If we have to save the Starchild's final one, okay, fair enough. But these tests must be as effective as possible, and tooth pulp is the easiest place for us to try to get recoverable nuclear DNA."

He intended to take two teeth from the adult skull's upper maxilla, one to test and one for me to keep safe for subsequent testing at another lab if they had positive results and confirmation became necessary. Then they would cut a square from the Starchild's upper left parietal to match the square taken from its right parietal by David Sweet for the same reason—its inner surface in that area could never have been touched by human fingers, and its upper surface of shellac could be sanded off to remove any contamination.

They arranged the removal area in a conference room so everything could be recorded by a digital camera. We wanted to leave no doubts about how the bone was harvested and prepared for testing because, if we obtained the results we anticipated, our many critics would promptly insist that we somehow had doctored the samples. More worrisome to me, though, was that a positive result for us would immediately put our critics in "kill the messenger" mode. I dreaded that.

The skulls were put side by side near a dark cloth spread on the table. Beside them was a powerful rotary tool that looked like a flashlight with a circular serrated blade attached to one end. That was a fresh 1-1/4" Dremel blade, the basic tool for taking bone samples. Four cuts would be made into the Starchild's upper left parietal, excising a one-inch square. That square would be cut into four segments, each as big as a nail on a little finger. They'd keep three for testing, and—as with the teeth—I'd keep the fourth for testing at another lab if success warranted confirmation.

They required much less bone than the BOLD Lab took, so I asked about that. For the first time, Jason and Ripan seemed uneasy. They brushed it off by saying techniques had improved greatly in the past several years, but I felt strongly that they were trying to avoid criticizing a peer.

The first order of business was to extract two of the adult's teeth, both right-side molars, which was done with pliers. Both required quite a bit of muscle from Jason because they were so solidly rooted into their maxilla.

When both teeth were secured in sterilized test tubes, Karena and I joined Jason and Ripan in putting on protective eyewear. Then Ripan continued filming while Jason put the Starchild skull on the cutting cloth so he could slice into it. The Dremel's high-pitched whine erupted and he pressed the whirling blade down to the skull's surface. *It bounced off!* Jason looked up through his protective face shield to glance at Ripan, who shrugged with confusion.

Jason went back at it, bearing down to hold the blade in position. This time it ground through the shellacked surface and bit into the bone, immediately filling the air with two things: powdered bone and the unmistakable stench of teeth being drilled into by a dentist. I didn't know it then—but later learned—that not just teeth but all manner of bone produces that distinctive odor, which is burning collagen, rather than enamel. A speck of flying bone thunked against my protective glasses as the smell overwhelmed us.

"Ohmigod!" Karena gasped, struggling to cover her mouth and nose with both hands. "I think I might gag!"

I felt much the same, and I could see Jason and Ripan struggling with it in their own ways until Jason managed the fourth cut that freed the square of bone, which then fell into the skull and out through its foramen magnum hole.

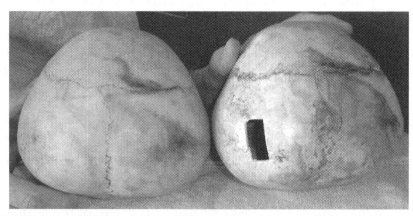

The SC copy at left, the real SC on right, showing window cut out by Jason Eshleman of Trace Genetics. The cut is more square than this photo angle makes it look.

"Whew!" Jason said from behind his face shield. "That was tough! I've never cut into bone so hard yet so thin!"

Colonel Philip Corso's description of the aliens recovered from the Roswell UFO crash echoed in my mind: "*The bones are thinner but seem stronger, as if the atoms are aligned differently for greater tensile strength.*"

"Can we open that window?" Karena rasped, pointing at it. "And the door, too . . . let's get a draft moving through here."

Ripan moved to the window, leaving the door for me.

"Extra-strong odor, too," said Jason, stating the obvious.

He retrieved the small square of bone and put it on a wooden cutting block, making two quick crosscuts to quarter it. More powder, more stink. "We'll cut our tooth when it's time to test them," Jason added as he worked. "No sense exposing its pulp to contamination until we have to."

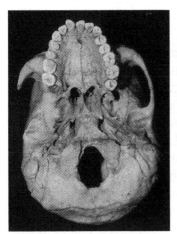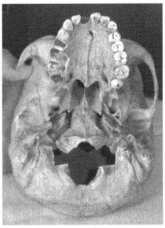

SFWS before (left) and after (right) samples had been removed for genetic testing. Note an incisor and a pre-molar have been extracted, but are not yet needed for testing.

For several minutes the room's air stayed laced with powdered bone and stink. As it gradually cleared, we speculated about how bone that looked so fragile could actually be so durable. I explained that one of David Sweet's assistants had told me they also found it difficult to cut in Vancouver. Now I realized he seriously understated that situation. Also, no one at the BOLD Lab ever mentioned the glaring increase in the smell of burning collagen. You'd think a lab specializing in dental analysis would have noticed that.

"It's resilient enough to be partially fossilized," Jason said, causing Ripan's dark eyebrows to lift. "But it's not. Here along the edge it's clear the matrix is as it should be. This is bone. Hard bone, unusual bone, but bone."

Later I learned the Starchild's bone has an extraordinary overload of collagen relative to its thickness (half of the human average). It was as if, I was told, a biological compression program was run on it, leaving its usual amount of collagen still in place, compacted into a space half as thick—and thus twice as pungent—as that size would dictate.

Standing in the conference room at Trace Genetics on the Davis campus, what mattered was that the bone was cut and sectioned, and in a few weeks we could have answers we'd tried to secure through four interminable years.

My heart soared like *Little Big Man*'s memorable hawk.

One skull fragment from the Starchild and one adult molar took their place in the testing queue at Trace Genetics. On March 7, 2003, the testing began. The molar was cut so that only the part buried in the maxilla would be tested, ensuring minimal contamination. However, the skull fragment had to be thoroughly decontaminated prior to DNA extraction.

The first step was sanding away its surfaces to remove oil from human fingers that may have seeped in, although given its high degree of hardness, seepage seemed unlikely. After sanding the fragment, it and the tooth were bathed in strong ultraviolet (UV) radiation for five minutes before a dip in sodium hypochlorite bleach. After that, they were rinsed and placed into test tubes filled with EDTA (ethylene diamine tetra-acetic acid) and sealed with parafilm before being placed on mechanical rockers. One week on those rockers would decalcify both bones and render their DNA usable. These were routine protocols to prepare ancient bone for testing.

Due to the press of other business, both bones were left in the EDTA solution for ten days. On March 17 they were checked, and the tooth was fully decalcified, while the fragment was unchanged. This was unexpected, but manageable.

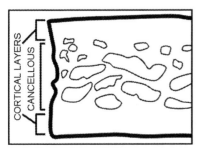

A digestive enzyme, protein-ase K, was added to soften the matrix by dissolving protein in the bone. The next day it looked as if some of the cancellous area (where marrow resides) had dissolved, though the outer surfaces seemed unchanged. It was even more durable than first assumed, but enough protein seemed dissolved to attempt an extraction.

As at the BOLD Lab, the adult's nuclear DNA (nuDNA) was recovered easily, verifying she was female. Her mitochondrial DNA (mtDNA) also was extracted with ease, which made sense because burial in a mine tunnel was like 900 years in a climate-controlled storage locker, with temperature and humidity naturally maintained at constant levels. Trace Genetics was able to place her mtDNA with one of the five major matrilineages (haplogroups that move only through females) of all modern full-blooded Native Americans. These are haplogroups A, B, C, D, and the rare X. The female found with the Starchild was from the haplogroup A, common throughout Mexico and the southwestern U.S.

Unfortunately, not enough protein was dissolved from the Starchild bone by the EDTA and the added proteinase K, so it produced a muddled reading that looked like a contamination. Assuming it was, Ripan and Jason could only try again, but this time they knew they were dealing with bone that couldn't be handled with protocols used for typical bone. They'd have to do several things quite differently.

Starchild went back in the queue and was retested beginning on April 21. The second fragment was sanded as before, but this time it was bombed with UV light for 15 minutes instead of the usual 5. Instead of the typical rinse with bleach, it was immersed for 5 minutes. Instead of 7 days in EDTA on a rocker, it was left for 22 days to be certain of sufficient decalcification so the proteinase K could dissolve the proteins in it. However, this still wasn't enough to dissolve the Starchild's astonishingly durable bone.

Realizing they needed unusual steps to create a thorough digestion, they added a strong detergent called Tween-20, which overnight dissolved the stubborn bone chip to nothing. Only a few flakes of its residue remained in the bottom of the test tube. "It was as close to complete digestion as I had seen from prehistoric bone," Jason later wrote in an email.

We held our collective breath as they proceeded to analyze it. Would its DNA be recoverable? Would it be contaminated enough to nullify any result? Would it be degraded by 900 years of burial and/or 60 in a cardboard box? That seemed unlikely given the ease of extracting the adult's DNA, but degradation would be a serious concern until we were sure both forms of the Starchild's DNA—mitochondrial and nuclear—were intact and recoverable.

We finally caught a major break. Well, half a major one. Jason and Ripan easily recovered mitochondrial DNA from the Starchild, *clean and uncontaminated!* In that regard, it was the best we could hope for. It meant the Starchild's DNA had not degraded to an appreciable extent, which in turn meant it was as well-preserved as the human found with it.

The analysis showed that its mother was indeed a human (which we expected because a female's egg is the primary engine in the creation of life—sperm go along for the ride). However, its human mother turned out to be from the Native American haplogroup C, another common type, though not as common as the adult female's haplogroup A.

On the surface, that would firmly establish that the person found with the Starchild could not be its natural mother. But that was only if the Starchild was conceived sexually. If the conception was in vitro, then she could have had the genetically altered egg of another female implanted into her own uterus, in which case she could have been the Starchild's birth mother without having a genetic connection to it.

An adoptive mother remained possible, or a devoted companion or caretaker willing to die with her charge, but the most likely explanation, at least from my standpoint, was that she was its "unnatural" mother. This provided a few more layers of intrigue to the death-and-suicide scenario, but mostly I was just glad to have this much answered.

What didn't come through was any hint of nuclear DNA, and most significantly, several attempts to amplify fragments of the amelogenin gene located on the X and Y chromosomes were uniformly unsuccessful. This meant that the faint microlevels (200 picograms of the usually necessary 1000) supposedly extracted *twice* by the BOLD Lab, were not found by Trace Genetics, a lab better equipped to recover and evaluate ancient DNA. Thus, BOLD's claim that the Starchild was a human male was, plainly and simply, a mistake.

Although my doubts about the BOLD Lab's result were now justified, it was much too late to recover any of the losses incurred. Damage had been done at too many levels. However, let me emphatically state that I think Dr. David Sweet and his BOLD Lab staff acted with proper professionalism and did what they thought was right and honest. They simply made errors while testing, errors they couldn't see or couldn't imagine. We all make mistakes.

In any case, the second round of tests at Trace Genetics told us an important part of what we needed to know about the Starchild's DNA—it was very likely no more degraded than the human found with it, which was barely degraded and easy to recover. Nonetheless, we still badly needed to recover and sequence the mystery skull's all-important nuclear DNA, so all they could do at Trace Genetics was knuckle down and once again try to secure it.

On June 4, a third round of testing began on the third of the four bone fragments originally cut from the Starchild skull on February 10. Again it was sanded and UV'd for 15 minutes, bleached for 5 minutes, and instead of the second test's 22 days in EDTA on the rocker, this time they left it 30 days. Again it was placed in proteinase K, to no appreciable effect, so again Tween-20 was added, which again reduced it to nothing overnight. Again the mitochondrial DNA was easily recovered, and again it showed the Starchild's mother was human and came from haplogroup C. This result was now beyond doubt, as was the lack of an overt biological connection between the Starchild and the person found with it.

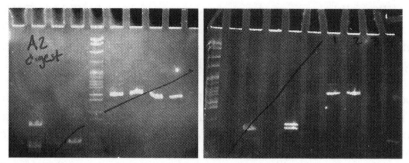

Vertical "ladder" is the nuDNA for an experimenter to show that the gel sheet works. Four bright horizontal bars on left sheet shows SFWS's mother was a human from haplogroup A. Four bright uneven bars on right sheet show SC's mother was from haplogroup C. Faint bars scattered about are called "primer dimer." They also indicate a gel sheet is working.

Unfortunately, during *six* attempts to amplify the Starchild's amelogenin gene, they never produced a polymerase chain reaction (PCR) product. Those six failures were documented in a pair of blurry, indistinct photographs of gel sheets that were clear enough to Jason and Ripan.

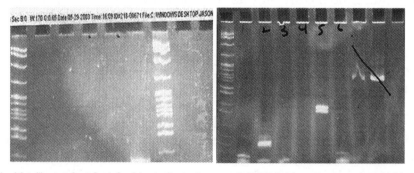

Ladder-like track at far left of both sheets is control. SFWS (left) produces a clear, ladder-like track of nuDNA on far right of sheet, revealing a female human. SC (right) shows only primer dimer in six attempts. If the Starchild's father was human, a track should be seen.

They were explaining the situation to Karena and me as we sat at an outdoor café table at the Student Union on the U.C. Davis campus. It was a beautiful summer day, and students milled everywhere around us.

"So what you're saying," I summed up for them, "is that once again you got a nice clear reading on its mitochondrial

DNA, but you're zero for six this time around at recovering its nuclear DNA. Right?"

Ripan nodded behind an uncomfortable shrug.

"How can that be?" I protested. "If its mitochondrial DNA is coming through loud and clear, we have to assume its nuclear DNA is sitting right there, ready to be recovered. Am I wrong to think that?"

"Not necessarily," Jason replied. "There is always the possibility of some kind of exotic degradation we can't understand or account for at the moment. Degradation has to remain the most likely explanation."

"That was the case with Kennewick Man," Ripan added. "Once its skeleton was exposed on the river bank, it was baked by direct sunlight and soaked and re-soaked by water, so its DNA was fully degraded. Basically, your skull's result is not unusual for us—it's more typical than not."

"Ohhhh, *please!*" I yelped. I was normally hard to agitate, but this was well over the top of absurdity and I could tell they knew it. "That skull was buried in a mine tunnel for 900 years! No sun! No water! *No degradation!*"

"Calm down, Lloyd," Karena put in. "It's not their fault."

Soft-spoken Ripan raised his voice a notch. "You're free to think whatever you want. If you choose to say, for example, that we're not recovering because primers aren't able to pick it up, that's your prerogative. But from a professional standpoint, we can't advocate that as an answer. Our position has to be that the result was due to degradation."

I understood what he meant. If the Starchild's nuclear DNA was not degraded, just sitting there waiting to be recovered, yet the primers weren't locating it, that could only mean one thing: *the nuclear DNA they were seeking was not entirely human.* Those primers were built to recover human-only DNA—not ape, not gorilla, fish or fowl—*human only!* That option was untenable for Jason and Ripan because it meant admitting something their scientific peers simply would not tolerate. Their careers, if not their lives, would be ruined if they stepped across that dangerous line.

"Okay," I said, forcing myself to simmer down, "I accept that for public consumption you two have to say the problem here is degradation. But I'm going on the assumption that human-only primers don't function at all on the Starchild's nuclear DNA. Do you guys have any problem with that?"

Ripan shrugged. "You can interpret it however you like."

He didn't have to paint more of a picture. "Okay, then, let's leave that for the moment. Where do we go from here?"

"We all have to wait," Jason replied. "The primers we use today have a degree of sensitivity that just doesn't work on what we have now, for whatever reason. However, in the not too distant future, say in the next several years, that sensitivity will increase greatly, to a point where it should begin to sequence at least some of the Starchild's nuclear DNA."

Several years!? The thought of waiting that long hammered me like a wrecking ball because I couldn't see where we'd gain any advantage with an improvement in primer efficiency. "Won't improved primers still be seeking human-only DNA? I mean, wouldn't anything they recover be from only the human part of a hybrid's genetic package?"

Jason and Ripan squirmed with discomfort. "I'm afraid that's true," Jason said. "It's just parsing your problem into a finer layer. You may never get the result you need."

"There's one avenue for hope," Ripan added. "It's a long shot at the moment, but who knows? Miracles can happen."

Karena and I glanced at each other, wondering how many more miracles—if any—we could reasonably expect.

"Everyone in our field," Ripan went on, "wants to sequence the DNA of a Neanderthal to compare to humans and chimps. That's the Holy Grail of ancient genetic research at the moment. But with human-only primers, those researchers face the same problem you face with the Starchild. Neanderthals were clearly not entirely human, so their nuclear DNA won't respond to any human primers. Another method has to be developed, and that's underway."

My hopes soared. "Really?"

"Some people are working on it," Jason put in. "We've heard about it on our professional grapevine. But it's years away from completion, *if* they ever figure out how to do it. Understand, the problem is finding a way to sequence every single base pair, all three billion! How do you manage that?"

I had tried to stay abreast of the sequencing of the human genome, so I had some grasp of the difficulty involved. "The people who do that kind of work . . . sometimes they *do* pull off miracles. I put nothing beyond their abilities."

"Neither do I," Ripan agreed. "So their research really does give you legitimate reason for hope. However, don't count on it. You need to accept that this problem won't be solved for at least three or four years, and maybe more, so you need to point your efforts in another direction."

"What I'd like to see," Jason put in, "is a full biochemical analysis of that bone, as much as your budget can afford."

Accepting that we'd hit a brick wall in one direction, I had no choice but to change course gracefully. "Sure, why not? That bone is too light, too hard, and it doesn't dissolve in typical bone solvents. The biochemistry of something so unusual can turn up any number of inexplicable aspects."

"Don't get your hopes up too high," Jason warned. "Any results you get in any scientific area other than DNA will be able to be explained *away*, if you know what I mean."

I knew exactly what he meant. Whenever scientists are presented with information or data they don't accept, they collectively stand behind their dogma by denigrating the new data, suggesting errors were made collecting it or interpreting it, but in no case could they be wrong about it.

That nearly always works for them until the new evidence piles up enough to overwhelm the old guard that doggedly fights to maintain their status quo. However, DNA evidence is much harder to "spin" in any direction. It says what it says, and it does so with compelling authority, even more authority than the longest list of academic credentials—individual or combined—can normally muster.

"Well, if it's three or four years before a break in DNA analysis, I have to do something with the bone. I'll definitely look into your biochemistry angle to see what I can find out."

"Do that," Jason said, as he and Ripan rose from the café table, indicating our meeting had gone on long enough.

Ripan reached out a hand to shake with us. "We'll let you know when we hear of any breakthroughs that can be applied to the Starchild's DNA."

"Any idea where I can go to get the biochemistry tested?"
"We can give you some names."

[See Appendix II for the full text of their genetics report.]

Karena and I left U.C. Davis in a stupor of disappointment. After rebounding so strongly in the first half of 2003, we now faced three or four—and perhaps several more—years of waiting for genetic technology to catch up with the Starchild's unique needs. Meanwhile, we had to find a way to arrange the biochemical testing Jason Eshleman and Ripan Malhi felt could provide compelling data. Unfortunately, the names they gave us were no bolder than most of the scientists I'd been trying to talk to. I found myself still confronting the same old story of too many "experts" fearful of stepping over the line separating science from pseudoscience.

Neither time nor temerity was on our side.

CHAPTER ELEVEN

CHANGE OF VENUE

The nature of the universe is such that ends can never justify means. On the contrary, the means always determine the end.
—Aldous Huxley

That last meeting with Jason and Ripan was in late July of 2003. By then a heady new player had emerged on the Starchild Project team, Grant Stapleton, a thirtyish South African yacht hand who'd immigrated to London ten years earlier. He was eager to take a role in various aspects of alternative knowledge, so I directed him to Belinda McKenzie and they were an instant match. He proposed a new cause to her, one designed to assist many alternative researchers in the same way I was being helped. That kind of expansive thinking was music to Belinda's aggressively proactive ears.

As she supported me, she began supporting Grant so he could focus on their project, called Cognoscence. She invited him and his partner, Eva, to move into the top-floor flat of her townhouse in Highgate, North London, to facilitate getting Cognoscence to fully function as soon as possible. Grant operated at the same frenetic pace Belinda maintained, so they meshed well. In addition, he gradually became an extension of me in London, learning the Starchild ins and outs as I instructed him, then taking off on his own to freelance around the scientific community, seeking out help for that cause.

Grant was a solidly built cherub of a man with reddish hair and a gift of gab like an Irishman. In fact, I suspected his South African ancestors must have been transplanted Irish. He could certainly talk the fleas off a dog, and soon he was hobnobbing by phone and by email with several of the top scientists in London within and outside the mainstream. Rupert Sheldrake, famous for his "Morphic Fields" theory, led to one connection, who led to another connection, who led to a scientist working at a place called Royal Holloway, a branch of the University of London. She was willing to help.

Grant Stapleton and new daughter Maya.

Let's call her "Jean." Jean invited Grant to provide her with a sample of the Starchild's bone so she could analyze it under Royal Holloway's high-tech scanning electron microscope. It would have to be done during off-hours and "off the books" so none of the university's pencil pushers would know its equipment was being used for something as absurd—if not unethical—as attempting to determine the genetic parentage of a bizarrely formed skull. I'm reluctant to admit we were forced to engage in such chicanery, but there it is, like it or not.

In late October, 2003, one year since Belinda first emailed me and changed my life with her support, Grant and Jean arranged to meet at the lab housing Royal Holloway's scanning electron microscope. He arrived with the fourth of the fingernail-sized fragments cut by Jason Eshleman from the Starchild's parietal back in early February at U.C. Davis.

It was mid-morning in New Orleans and mid-afternoon in England when my desk phone rang. "Hello?"

Grant's accented words blurted out. "Lloyd! Grant! I'm just back from Royal Holloway with incredible news!"

His excitement instantly infected me. "What is it?"

"You know the cancellous holes between the bone?"

In flat bone, like in a skull, cancellous holes were sandwiched between the upper and lower cortical layers. In life, cancellous holes held the marrow that manufactured blood cells. After death, they were scoured clean by bacteria.

"Yeesssss," I hissed, expecting something momentous.

"The cancellous holes have some kind of weird *fibers* extending from them, and they seem to be a part of the bone!"

"*What?*"

"Fibers, mate—crazy, weird fibers! Jean's never seen anything like them before, not in human bone or any other kind of bone. She thinks they could be absolutely unique!"

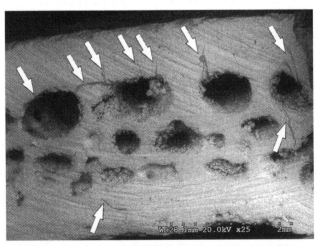

Fibrous strands (indicated by arrows) embedded in the matrix of the SC bone.

"Ohhh, geeze . . . this is *huge!*"

"Yes, and it gets better! We also found extraordinarily high levels of aluminum all through it, saturating it!"

"*Aluminum!*" At high levels, aluminum was poison to tissues. I knew that much, and that it was supposedly a contributing factor in Alzheimer's. High levels in bone I wasn't as sure about. "What the heck does *that* mean?"

"We don't know, but we have photos of the bone scans and chemistry scans. As soon as they're processed and available, I'll forward them to you. Let's face it, though—this changes everything. Now, what's next?"

I had to struggle to stay reasonably calm and collected. "I don't know, Grant. This is really a bolt from the blue. I think I need some time to think about it."

"Belinda and I have talked it over. We don't think we can handle this ourselves. Nobody can talk about the Starchild the way you do. Nobody knows it like you do. We want you to think about moving over here to live full-time until we can figure it all out. Would you consider doing that?"

"Moving to London?"

"You can live here in the house with us in an empty bedroom. And if you do come, we can marshal our forces into a truly effective team. We really do want you to come."

"Geeze, Grant, that'll cause a lot of turmoil in my life."

"We don't feel we can handle it properly without you."

I held a long pause, then said, "I'll see what I can do."

The next day an email from Grant appeared in my inbox, loaded with attachments. There was no explanation of what I was looking at in any of the shots, just the shots themselves. Still, I could see—anyone could see—that the fibers emerging from the cancellous holes were highly unusual, and the graphs showing high aluminum spikes were equally bizarre. Comparing them to the normal graphs Grant sent with them was a lesson in itself. However, it was the black-and-white photos of the fibers that really did it for me.

Those fibers were undeniable real, they showed a significant range of variation, and their edges were shredded, having the durability to resist a whirling cutting blade. They were simply astonishing and, to my knowledge, totally unprecedented in any bone. Those fibers hooked me hard and left me with no doubts about moving to London. Now I *had* to go.

Three graphs showing aluminum (Al) levels in bone. Normal bone (left) and two elevated samples from SC (center, and right). All major minerals are listed by their chemical abbreviations.

Scanning electron microscope (SEM) images of SC bone. Top image's squares are around two different kinds of "fibers" seen individually under higher magnification.

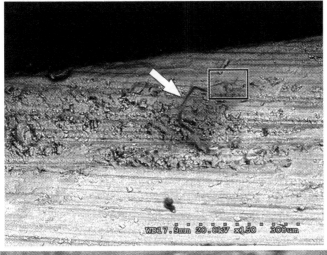

SEM image of fiber apparrently cut loose from the bone and snagged on the surface. Note that the fiber was so durable that a high speed blade did not cut it cleanly.

It's difficult to pull up stakes after six years in the same place, doing the same things, whether it's a job, a home, or a lifestyle—and usually it's all three. Predictability can become a rut, but a rut can be comfortable, and you'll often resist leaving it. I did, though I had other reasons as well. Living nearby were my aged parents, an infirm aunt, and a brother battling severe diabetes. The worst could happen at any time while I lived in England, so any extended separation would risk that I'd never see one or more of them again.

Karena didn't want me to go for the obvious reason that San Francisco to London could stretch our long-distance relationship badly out of shape. She could afford flying to New Orleans every few weeks, but after I moved, the round-trip cost would soar from $200 to $2000—impossible on a regular basis. Her personal circumstances dictated it would be best if I flew back to the States to visit her every three or four months, which I was willing to do, but I couldn't stay for more than a week or so, at most. Work had to come first.

We tried to stay positive, but we weren't kidding ourselves. We both knew this was likely to mean real trouble for us.

On the other hand, Grant and Belinda were right to think everything regarding the Starchild had changed. The news about those fibers woven through the bone and the aluminum permeating it seemed to be what I had struggled toward since 1999. It felt like a gift from the powers that be, as much a gift as Belinda's wholehearted support, and now the gift was being extended in ways and for reasons that seemed almost predestined. I felt, rightly or wrongly, that fate was leading me along this path and I could only follow.

I stayed in Louisiana through the end of 2003 to spend what might be my last holiday season with one or more of my ailing family members. In January, I gave notice on my apartment and started putting my belongings into storage. My commitment to Belinda was open-ended, so I planned to stay in London as long as I could learn useful things about the Starchild skull. Grant Stapleton insisted that English scientists were not as intimidated as their U.S. counterparts by pressure from peers to toe the mainstream line.

Because we consistently faced such strong resistance from scientists in the U.S. and Canada, I found it hard to believe they behaved differently in England. However, I was willing to give Grant a provisional benefit of my doubt, meaning I would accept his view when I saw it for myself. Even under ideal circumstances, though, I'd need to stay in London from several weeks to several months. I planned to stay for up to a year.

I arrived in late February of 2004, which was coincidentally the same week, five years earlier, that I had first laid eyes on Ray and Melanie and the Starchild skull. It seemed a fortuitous beginning to the journey. However, arriving in London in the dead of winter can be disconcerting to a Southerner like me. All of the stereotypes about London's dismal weather exist for good reason. Also, I picked up a flu bug on the flight over, so I spent my first three weeks there sticking mostly to my new room, being sick and fighting the "chilly damp."

Not so fortuitous after all.

Belinda McKenzie was a female Don Quixote, a handsome woman of my age (late fifties) with a mane of thick brown hair framing a face chiseled lean by years of jousting with social, moral, and ethical windmills. Not only was she bankrolling me and Grant and the Starchild effort, she was one of the leaders of a group determined to change England's antiquated charity laws. She also lent her support to a wide range of other issues and causes in England and Paris, the home of her own long-distance partner in life, Moj, an intellectual poet with whom she spent her free weekends.

Belinda was a dynamo of activity supporting her passions and crusades, and I soon realized—as did Grant before me—that the best course was to not question or wonder about her many other interests, but to just be grateful she had taken the two of us under her courageous wings.

We lived in comfort in the Highgate townhouse: Grant and Eva up in the third-floor flat; Belinda and I and occasionally her daughter, Tizzy, in three bedrooms on the second floor; and a ground floor devoted to Belinda's always busy office, a TV room, a dining room, and a kitchen. And that was not to mention the absolutely smashing garden, which seemed to be a staple of every dwelling in our neighborhood.

After living alone for so long, I found my new digs a bit hectic at times. However, I sure couldn't beat the price, and the frequent swirls of activity were always invigorating.

Belinda McKenzie

Because of my lingering bug, Grant and I didn't meet Jean at Royal Holloway until late in March. It is a stunning place that looks like a medieval castle, with turrets at the corners of walls and crenellations along the tops. It was one of the most majestic schools I had ever visited, yet we met Jean in its very modern, very ordinary student union building.

Jean was a charming, good-natured biologist in her late thirties who first struck me as a hippie gone straight. However, the ease of her initial chitchat soon gave way to what made her what she was. She settled down to business and didn't mince words—technical or otherwise—as she led me carefully through the meanings of the first round of analysis by the scanning electron microscope (SEM).

"The bottom line," she concluded, "is that the SEM indicates you have a genuine anomaly. To be scientifically valid, it should be handled in step-by-step fashion, with the first step being a try at ruling out any kind of biological contamination that might have created the fibers postmortem."

"What kind of biological contamination?"

"Bacteria, fungi, yeast, molds. . . ." Grant replied. He had already been through it with her and knew the drill by heart.

I turned back to Jean. "Who can rule that out for us?"

"Mycologists," she replied. "Some of the best in England work not far from here. Mike Rogers is one. I can help to arrange a meeting with him if you like."

"Absolutely!"

Mycologists, it turned out, specialize in studying fungi. Jean arranged for Mike Rogers to evaluate what we had, so I busied myself turning photos taken by the SEM into a visually appealing Power Point slide presentation.

The following series of black and white images was taken by a scanning electron microscope (SEM). Each image is taken at higher magnification than the preceeding one. We refer to it as the "claw" series. The flakes on the strands are bone dust from cutting it.

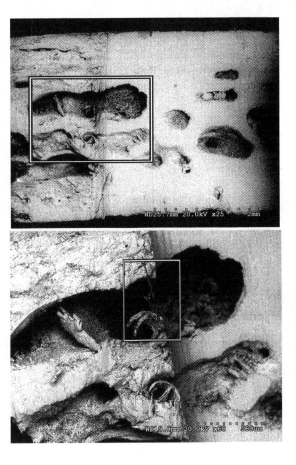

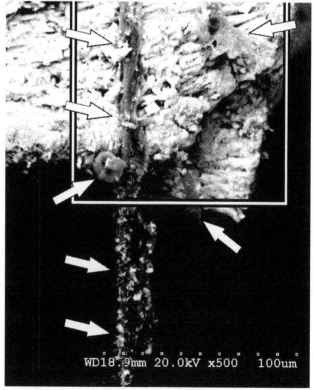

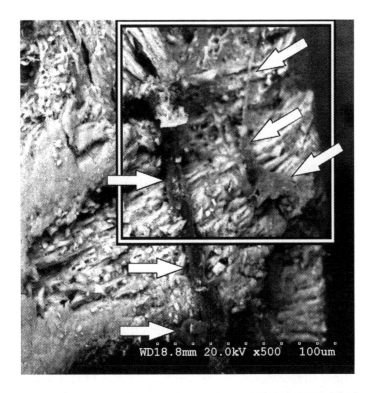

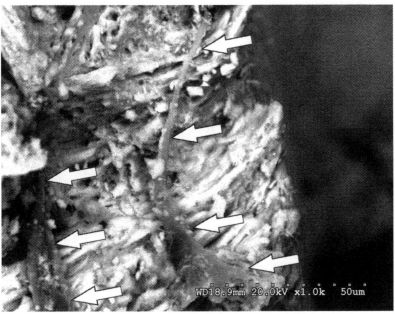

Mike Rogers was a polite, genial young man who, after seeing the first few slides, went into his building to round up a pair of colleagues. It clearly wasn't every day that something like this came through their doors.

The three men studied each slide, asking numerous questions, but off the top of their heads they were unanimous that they couldn't recognize the Starchild fibers as any kind of growth. However, they had a small caveat.

"We have *a-boot* 30,000 possibilities to consider," one of them said, with a distinct Scottish lilt. "So what we *think*, though we're confident of it, matters not a whit. You'll be needin' a Maldi-Tof test to be certain."

"Where can we get one?" I asked.

Mike Rogers smiled. "We can help you arrange it."

The Maldi-Tof test turned out to be very complex, time-consuming, and expensive. Belinda had made it clear we could have money to do what most needed to be done, but her funds were limited and we had to use good judgment about how we spent them. Therefore, Grant and I put the Maldi-Tof test on a back burner because we could see the mycologists knew their business. They seemed quite confident the fibers were an integral part of the bone rather than something grown onto it or into it after the Starchild died.

We also decided against it because Colonel Corso's assertions kept rattling around in our heads. Whether he was genuine or a fraud, in his book he had accurately described the Starchild's body (*about four feet tall*) and its bone (*thinner but stronger*) one year before it was handed over to the Youngs. That persuaded me to believe if something had to account for the durability of the bone, it probably was those odd fibers.

Our next step was to bring in another expert with a wider range of forensic expertise than Jean possessed. She knew a man, she told us, who owned an independent forensics lab dedicated to the study of soils and other inorganic materials found at crime scenes. He was a forensic geologist, but he was connected to all manner of other forensic specialists.

Jean believed that if he agreed to get involved, it would be a huge step forward for us.

"What's his name?" I asked.

"Ken Pye," she replied in her usual blunt manner.

I thought I'd misheard her. "Ken *what?*"

"Ken Pye," she repeated, cracking a smile. "No relation."

"You're kidding!" I blurted. "You must be!"

Her head shook, causing her heavy single braid to swish against her neck like a pendulum. "No, that really is his name. But do understand that here in England the name *Pye* is not so rare as in the States."

That was true. Nearly all of the Pye family's ancestors came from the United Kingdom of England, Northern Ireland, Scotland, and Wales—mostly Scots and Welsh.

[Aside: We have a poet laureate of England in our background, Henry Pye, widely disdained for an overabundant use of bird imagery, among other flaws of style and character. From his tenure, in fact, came a mocking refrain sung by children in the streets of London, which was how all social protest was lodged against officialdom when literacy was uncommon and a wrong word at the wrong time by a disgruntled adult could mean a sojourn in the stocks—or worse.

Sing a song for six pence and pocket full of rye,
Four-and-twenty black birds baked by a Pye.

The modern version has changed slightly in wording and vastly in meaning, as have most such nursery rhymes, the majority of which began as the "protest songs" of their era, crafted by poetic wits to be sung aloud by those who could do so in public with collective impunity.]

Dr. Ken Pye agreed to meet us for lunch at a restaurant near his office. He turned out to be an average-sized, tightly coiled man in his mid-40s, with close-cropped, neatly combed hair indicating the orderly mind he had to possess to do what he did. He was serious and didn't waste smiles, but when he let one go, he lit up the room. I liked him right away and so did Grant. He generated an air of supreme confidence and competence, both of which we needed in abundance to proceed forward from where we were.

A somewhat younger Dr. Ken Pye on expedition in Australia.

It didn't hurt that he and I shared our surname, but it also didn't hurt that he owned his own laboratory, which meant that, unlike Jean, he didn't have to be concerned with the disapproval of straitlaced colleagues. He could do whatever he wanted, when and how he wanted to do it.

After lunch we joined him at his business offices, where he moved us into a conference room with a long, polished table. He indicated we should sit opposite him, which we did as I unzipped my carryall to pull out the stereolithographic reproduction of the Starchild, plus the maxilla half.

[Reminder: Ray and Melanie Young keep both skulls in El Paso, while I carry the copy and the real maxilla piece.]

I explained the skull's salient features, then handed it over to him. He held it firmly, twisting and turning it, hefting it, bonding with it and the half of the maxilla. Then Grant pulled out the photographic evidence from Royal Holloway's SEM analysis, which we'd discussed extensively at lunch. Grant went over those while Ken still held the skull, occasionally glancing from the photos back to it as he tried to absorb this ton of new information in one go.

Finally, he looked up and spoke. "I have to agree with Jean. I think you may have something completely unique here. I'd certainly like to know what it is."

"Does that mean you'll help us?"

He nodded, smiling faintly. "I'll see what I can do."

Grant and I could barely contain our elation. "What do you want?" I asked. "What do you need? How can we help?"

"I'm afraid I need fresh samples from both skulls, bigger than this little chip." He pointed to the small piece in the test tube that had been analyzed by the SEM. "I'll need to run an array of tests on them."

"Do you have a preference for where the samples come from on the skulls?"

"It's best if they come from the same area on both, for side by side comparisons. If I have a choice, I'd like it be from the parietals. Smooth surfaces, easy to work with."

That meant cutting into the adult's cranium for the first time and the Starchild for a third. "All right, I'll tell the owner, Ray Young. He'll do the cutting. And I'll make sure he only widens the square that's already in it.

"Ask him to use a fresh blade coated in diamond dust."

Diamond dust! That sounded seriously expensive. "Gee, Ken, I don't know. What does something like that cost?"

"It's artificial diamond, not real. Only doubles the cost."

"May I ask why that's important?" Grant put in.

Ken pointed to the SEM element graphs. "These aluminum readings aren't consistent. If it permeates the bone, they should be. Also, the Dremel company coats many of their blades with aluminum sulfate, so I suspect that's what the SEM picked up—blade contamination."

Grant and I looked at each other with abject dismay, both of us thinking, *Jean!*

"Why wouldn't Jean," I wondered aloud, "be aware of something as important as that?"

"Inorganic analysis isn't her field," Ken said. "It's mine."

I lifted my hands in a gesture of surrender. "Thank goodness for that! Let's just do whatever it takes to get this underway, resolved, and finished."

Ken leaned forward on his side of the table. "I have to ask this, Lloyd. I can see how committed you are, how keen both of you are. But what if, at the end of the day, we find your skull is merely a unique deformity?"

I shrugged. "If it is, it is. We'll accept that and move on. But living like this, in an endless limbo . . . this is hell. I want to punch my ticket and get off the ride."

One of his rare smiles flickered. "Keep in mind, I maintain quite a busy schedule. Other cases stand in queue ahead of yours. However, I'm personally interested in this one, so I'll get to it as soon as I can."

That was always how it seemed to be: hurry up to wait.

In early May I received an email from Andrea Cross, a student from Canada studying for her degree in Forensics and Investigative Science at a university in northern England. She told me she was doing an undergraduate dissertation on unusual anthropological relics and was interested in including the Starchild skull as part of her lineup. She asked, rather tentatively, if I'd be willing to meet with her and show her whatever material was available.

I explained that the original skull stayed with its owners now because it was too cut up to be on public display. I softened that disappointment by adding that she was welcome to come assess the stereolithographic copy, as accurate as modern technology could make it, and view the real piece of maxilla, which showed how its bone looked and felt. She agreed to come to Highgate to examine what I had.

In our exchanges I detected skepticism, but she was only behaving like a student in any scientific field. She was buying into the mainstream viewpoint, so she had no choice but to join the team full-bore, with no reservations. Besides, I was just killing time waiting to hear from Ken Pye, so her visit was something interesting to do. I looked forward to it.

Andrea Cross turned out to be a "modern" young lady in every sense. Dark hair streaked in electric shades of fuchsia and magenta. Several visible piercings and, I assumed, others not so visible. But within minutes her appearance was a non-issue as Grant and I settled in to deal with a topnotch intellect. She turned out to be whip smart and chock-full of moxie. She batted questions at us and we returned each, and as the match wore on, her entire demeanor changed. She hunkered down with everything—the idea of it, the actuality of it, the *potential* of it. She simply got hooked.

Andrea Cross

After three hours, as she packed up to catch her train back north, this ex-Calgary girl said her good-byes in the same lilt as Chad Deetken and Ted Robinson. "I think I'll make the Starchild the *only* subject of my paper."

That was surprising. "Can you do that?" Grant asked.

She gave us both a pointed stare. "Why not?"

"I don't think your professors will allow it," I replied. "I mean, you're lucky they're letting you include a thing as *verboten* as this as even a part of what you're doing. I can't imagine them letting it become the whole."

"They haven't seen it like I have. I'll tell them that based on my initial analysis, I think it's worth investigating."

"You think they'll believe you?" Grant asked.

"They won't think I'm lying, I can tell you that!"

I looked at Grant, he looked at me, our eyebrows lifted with appreciation for her verve, then I turned to Andrea. "Let us know if there's any other way we can help you."

Ken Pye, we discovered, flew all over the world to testify as an expert witness in court cases where his expertise was required. So when he said he maintained a busy schedule, he meant it. Apart from occasional brief chats on the phone, we didn't hear from him until late May, when he called us to arrange a meeting the following week. He said he had some significant news to share with us.

The meeting took place in his office, in the same conference room as before, sitting across the same conference table. This time he was noticeably more formal, creaking with propriety as we endured the required chitchat.

"All that aside," he finally said, "let me tell you why I asked you over. First, I want to assure you the fibers are real. No question about it. We have no idea what they are, but they do exist and they do present a mystery."

Grant and I nodded. How could he conclude otherwise?

"However, I have bad news, too—the aluminum."

Everyone around the world, it seemed, except Americans, spelled and pronounced that word as *al-u-min-i-um*, which always gave it a 'wrong' sound in my ears.

"It was indeed a contamination from the original cutting blade. So half of what you thought the SEM found was a mistake. I'm sorry about that."

As soon as Ken first mentioned coatings on Dremel blades, Grant and I had accepted that the aluminum reading was most likely an honest mistake.

"Understood," Grant said. "We were expecting that."

Ken's features hardened even more. "Glad to hear it. Unfortunately, I have a bit more bad news for you, and this, I'm afraid, is quite serious."

We sat back in our chairs as Ken leaned sideways to lift a file folder from the briefcase he'd brought in. He opened the folder to reveal a couple dozen color photos, a big change from the SEM's stolid black-and-whites.

He spread them across the table in a specific order.

"We did these here, using our binocular microscopes," he said as he arranged them. "I wanted to have a look at the bone in the circumstances we usually work with, which is much different from what the SEM can do. We took the two samples you provided—the ones cut with the diamond-coated blade—and we broke off a small piece from each."

"Broke off?" I repeated.

He nodded. "Normally, when you cut bone with a blade, the cancellous holes get coated with powdered bone. The degree of coating depends on the quality of the blade used."

Ken pulled from his briefcase a pair of black-and-white shots of the powder-coated cancellous holes cut by the first blade and taken by the SEM.

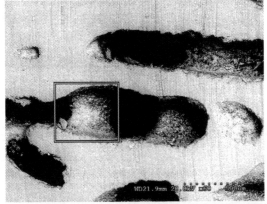

Cancellous holes are porous sections of bone between the outer cortical layers, through which marrow and blood passes. Rectangle shows a magnified view of bone dust sprayed into cancellous holes during cutting of samples.

"What you see in these shots is quite typical for cutting with a standard Dremel blade. Because of the density of normal bone, it produces a heavy encrustation of powder."

Alongside those he placed two color photos of the much smoother cuts made by the diamond-encrusted blade. "Now compare these and see the difference. Like night and day."

"How does a diamond coating reduce the powder so much?" Grant asked. "I mean, isn't a blade more or less a blade? Don't they all cut through the bone the same way?"

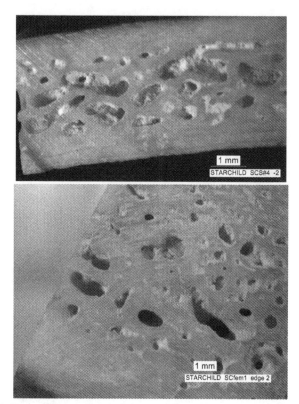

Bone cut with diamond covered blade. SC at top, SFWS at bottom. (Photographs are taken at the same scale; however, since SC's bone is half normal thickness, it takes up half as much lens space as the normal human bone of SFWS).

"It creates a much finer cut," Ken replied. "But then we can avoid any coating at all if we use pliers to snap chips off of one corner of each sample. We then carefully sand down the chip edges with lapidary tools, the kind jewelers use on precious stones. In normal human bone, that brings out a surface veneer that resembles certain kinds of alabaster."

The female's bone did look like alabaster, and its cancellous holes were clear enough to almost twinkle in the light.

"So clean!" Grant muttered. "You could eat out of them!"

"Something did," Ken countered. "Our living bodies contain certain parasites and bacteria that activate when we die. Their job is to help consume us down to our bones, then molds, fungus, and acids in soils take over to digest those."

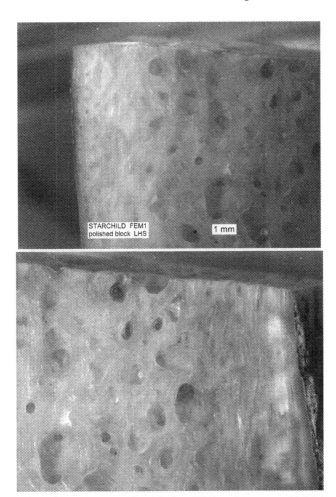

The SFWS bone "snapped" and polished by jewellers' tools to resemble alabaster. Note the extreme cleanliness of its cancellous holes, scoured out by bacteria after death.

Grant and I shared a look, then Ken continued. "What you see here is a typical result of the internal factors at work over a period of time. They devour every microgram of marrow from the cancellous holes, leaving a clean plate, so to speak."

We nodded, awestruck by the grim efficiency of nature.

"However," Ken went on, "when we did the same with the Starchild, this is what we found. Notice any difference?"

Grant and I bent over the photo he shoved toward us.

"It's milkier looking," Grant replied. "Not like alabaster."

"Right," Ken said. "That's apparently caused by all the extra collagen it has. We make that assumption because we've
never seen bone polish up to this degree of opacity."

He paused, urging us to try to locate what he wanted us to
find. "Check the center of the photo . . . dead center."

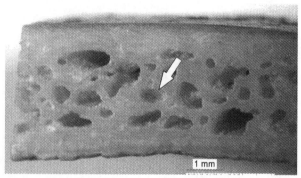

Even after "snap and polish," the SC bone is milky due to extra collagen. Note that a dark
(red) spot (difficult to see in black and white) is visible in the cancellous hole.

In one hole was a speck of something with a reddish hue.
"That little red spot?" I asked, hesitantly.

"How about this?" He slid another, clearer one to us.

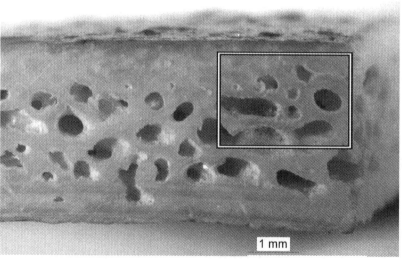

Multiple examples of reddish residue in SC cancellous holes.

"Whoa!" Grant yelped. "Red specks all over the place!"

"Here, too," Ken said, sliding another our way.

No doubt about it—those red specks were abundant.

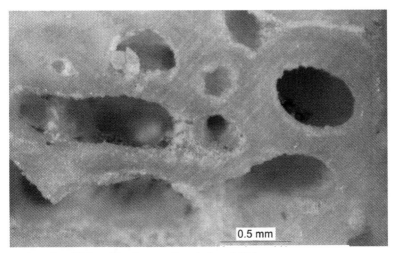

Closer view of reddish residue in cancellous holes of SC bone.

"Do you realize what those mean?" Ken asked, as somber and serious as Grant and I had ever heard him speak.

We sat back in our chairs, bracing for the shock of it.

"Gentlemen, I wish I knew how to break this to you gently, but . . . I think you've been hoaxed."

My heart thudded down to somewhere around my knees. *Hoaxed! How? Not after so much time and so much testing!*

Grant and I were both struck speechless. He stared at me, bug-eyed, while I had to be pasty white from shock.

Needing to finish what he started, Ken continued with his verdict. "The red residue we discovered indicates that the Starchild skull is a recent death—not ancient at all."

Grant looked as if he might faint. I think I almost did.

After receiving the skulls in 1999, one of the first things I arranged was a Carbon 14 test to find out how old they were when they died. That test required a substantial bone sample, but I didn't want to cut into either skull until I absolutely had to. Luckily, the adult skull had thin sheets of bone in its nasal opening, which could be removed without damaging its appearance. Since they supposedly died at the same time, one's date should hold for both. The Carbon 14 result was death at 900 years ago, plus or minus 40 years.

"But I . . . I had them carbon dated!" I managed to gasp out through my stunned disbelief. "How could it be wrong?"

"You told me you C-14'd the adult skull only, so I accept that *it* died 900 years ago. But if the Starchild was equally old, it shouldn't have visible residue of *any* kind in its cancellous holes. A recent death becomes far more likely."

"How recent?" Grant wondered.

"Depends on conditions of burial. Damp ground or dry ground; with coffin or without coffin; acidic soil, alkaline soil. I could guess and say this one looks like no more than a few years outside a coffin—but please don't count on it."

A few years! My mind reeled at how far I had been led from the truth . . . or had I led myself?

"What's that red stuff?" Grant put in. "Marrow?"

Ken shrugged uneasily. "That's the one bit of the puzzle we're still not clear on. Desiccated marrow is usually quite dark. I'm sure you've seen dried blood; it's almost black."

"Then how could it stay red?" Grant pressed.

That was a good question, and the moment Grant asked it, I could feel air seep back into my lungs. I looked at the photographs one more time—definitely red. I could feel myself calming down, starting to breathe properly again.

"Yeah! What *about* that?" I joined in.

"It's definitely an anomaly," Ken agreed, "but it's something to concern ourselves with a bit further down the line. For now, the primary objective must be a Carbon 14 date for the Starchild. If its death *was* recent, then your discovery story isn't true, and if *that* isn't true, we have to consider the unfortunate notion of a fraud or a hoax of some kind."

"But . . . Ken," I spluttered, "*those fibers!* Who would go to all the trouble to manufacture and insert such miniscule details *into the bone?*"

"Or make the bone so thin and light?" Grant added.

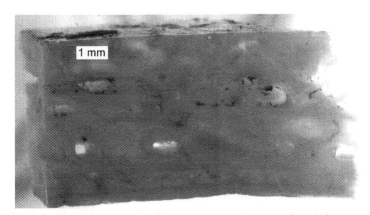

Back-lit sample of SC bone with abundant reddish residue in cancellous holes.

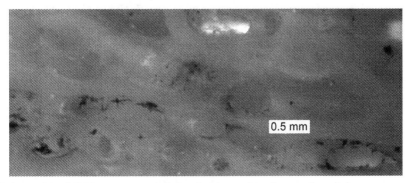

A closer view of the bone segment above showing residue in cancellous holes.

"I'm not suggesting it was manufactured. That's not possible. It's clearly a natural skull. I'm saying the story you were told might be concocted to hide a recent death."

"Okay," I said, "let's assume it *was* recent. Let's say it fell out of the sky and landed on this table. Would that change any of its biochemistry?"

"No," Ken admitted. "It remains a very unusual relic. But when you start with a lie, it gets difficult to explain test results in terms that scientists everywhere will be comfortable with. It seems plotted from the outset."

I had already rebounded from a fair share of disappointments along this journey, but I couldn't imagine how I would rebound from this one.

Grant didn't seem to be taking the news as hard as I was, but then he didn't have years of his life invested in it, either. He waited until we were away from Ken's office and driving home before he let me in on why.

"You know what went through my mind when he said what he said about it being recent?"

"An avalanche went through mine."

"Roswell!"

Due to the writings of Colonel Corso and dozens of others like him, all serious UFOlogists knew the Roswell story's main points. In July of 1947, at least one and possibly two UFOs crashed in isolated back country in the general area of Roswell, New Mexico. A nearby Army base mobilized to retrieve the craft and its passengers, and then they clamped an extremely tight lid on all information—public and official—about the matter. A man named Jesse Marcel was at the heart of the controversy, claiming he retrieved pieces of one craft to show to his family, which his son, Jesse Jr., had consistently and strongly verified in his adult lifetime.

Jesse Marcel retired to Houma, Louisiana, where my father grew up and where I was born. I knew people in Houma who knew Marcel before he died, in 1986, and they were convinced he wasn't the kind of man to make up a story that would create such widespread, vitriolic controversy, and then stick to it through all of the skepticism and harassment he had to endure afterward. They insisted that if he could have stayed well away from it, he would have, because he didn't seek the limelight regarding his experience.

That, plus the myriad facts of the case, led me to believe Roswell was indeed an actual event, and the truth about it had always been suppressed by our government.

"Roswell?" I repeated to Grant. "Why Roswell?

"What if the skull belonged to one of the Roswell aliens?" Grant blurted, excited by that possibility. "What if it's not as old as you were told, but it's older than Ken thinks?"

It was such an improbable, off-the-wall idea, I momentarily considered it. "Why would my government ever turn one loose? What could they possibly hope to gain?"

"Maybe they didn't turn it loose, maybe it was stolen!"

"*Stolen!* By who? By what? Any alien bodies those people control would be guarded like Fort Knox! They could be *in* Fort Knox, for all we know! This is a wacko idea!"

"Not necessarily!" Grant countered. "What if someone on the inside decided they wanted the truth about UFOs and aliens to come out? That could happen!"

"*Ha!* Not in the real world as *we* know it!"

"Would you just consider it for a minute?" he insisted.

I tossed it around for a few miles, then said, "Okay, assume it *was* stolen from inside the government system, from deep in the blackest core of the basement. Wouldn't they have taken it back while I was in Laughlin? Or as soon as they could get organized to hijack it from me?"

"Not if they *wanted* you to have it."

"C'mon!" I laughed, thinking he was joking. "Be serious!"

"I am!" he protested. "Maybe they've decided the time is right to finally come clean about UFO reality, so they figured this was a good way to do it. Release one of the skulls they have, give it to somebody in the field of UFO research, and let that person slowly bring it to public awareness."

"*Slow?* They sure made a mistake with me! *I'm glacial!*"

He shrugged. "Maybe this is what they expected of you. Maybe they knew it would be a years-long grind, and you're doing exactly what they want."

"That is nuts! They couldn't know how inept I'd be! Hell, *I* didn't know how inept I'd be! What if they wanted someone to do a *good* job of getting the word out? You think they'd let *me* keep it all these years?"

"How could they take it away? Everyone would know!"

The idea of the U.S. government sitting on its hands in frustration because they picked the wrong guy to introduce aliens to the world . . . I couldn't help laughing at that.

Grant smirked at me. "If the Starchild's Carbon 14 result comes in at sixty years ago, you won't be laughing."

He was right. If we were holding one of the Roswell aliens, we were ignorant pawns in a vastly larger conspiracy than we—or Ken Pye—could imagine.

That dismal thought chilled the car into silence.

CHAPTER TWELVE

TEST OF NERVES

*Only two things are certain: the universe and human stupidity;
and I'm not certain about the universe.*

—Albert Einstein

A piece of the Starchild's skull was shipped to a lab in Miami that specialized in dating bone by Carbon 14 analysis. Much is made of the method's potential for inaccuracy beyond 10,000 years, but anything within that is usually solid. We felt sure we could count on its results.

The dating sample was mailed on June 3, with a promise of a 30-day turnaround. Now all we could do was wait, sweating it out, as a major crisis developed in the house with Belinda. A critical tide was turning.

When Grant and I told Belinda the results of our meeting with Ken, some wind went out of her sails. She closed the door to her office so no one else in the house could eavesdrop, then she let us in on something we never asked about and she never discussed—her financial situation.

We knew she was born into a wealthy family, and she had inherited a substantial sum when her father passed away. She set aside a large part of it for each of her four children, then used her own portion to begin supporting the social, political, and now scientific causes that struck her as worthwhile. She took into her life and cared for stray people and noble causes the way others might take in stray animals, and that habit was costing her dearly.

"At the rate we're going," she informed us, "we have until the end of this year, then I'm afraid I have to pull the plug on everything. You need to know this so you can arrange your lives accordingly. I'm doing the same, and I don't want either of you caught unaware. This is a serious situation."

By then Grant and I knew Belinda's style quite well, and no matter how bad she made something sound, under the surface it was much worse. Hers was the proper British manner—understate everything and keep a stiff upper lip.

Grant chose to see this new dark glass as half empty, which no one could blame him for doing. Weighing it all out with his partner, Eva, they decided to cut his looming losses early and moved from the upstairs flat as soon as they could arrange a fresh start in another part of London. We could only wish them well as I moved from my room, the size of my bedroom in New Orleans, to the flat, the size of my apartment.

With that move into the top-floor flat, I became far more comfortable physically, but losing my *amigo*, Grant, left me emotionally quite bereft. He had shown me the advantages of a full-time partner, so I no longer preferred working alone.

While Grant and Eva packed up and moved, I made a long-planned and already-paid-for trip to the States for a week to attend my 40-year high school reunion. I grew up in Amite, Louisiana, a small town notable nowadays for being fifteen miles south of Kentwood, the hometown of Britney Spears. It is a conservative part of a Red State, so my "unusual" career made me the moral equivalent of a Muslim jihadist or a gay hairdresser, with no odds given on which was considered worse by the majority of its citizens.

Our 1964 class of eighty students had already held three reunions—after 10 years, 20 years, and 30 years. At the first, when we were 28, my partner was 38, which created a scandal of sorts. At the second, when we were 38, my partner was a cosmopolitan architect from Portugal who went braless in a see-through blouse—another scandal. At 48 I went alone, cause for another small buzz of disapproval. Now, at 58, I'd bring Karena, another woman of 38 whose flashy personal style would create yet another scandal.

From 38 to 38 in 30 years. I think that's what Einstein meant by relativity.

Sure enough, Karena was the youngest "partner" there, causing the expected bevy of wagging tongues and shaking heads. Feeling the heat, she asked me to not be truthful when my turn came to stand up and tell my classmates what was happening in my life. But I just couldn't resist.

"I'm living in London now, in a rooming house of sorts, doing my best to prove that a strange skull that's come into my care is actually from a human-alien hybrid."

If Britney happened to be in Kentwood that evening, she could have heard the collective gasp of disapproval. And I have no idea what Einstein would have called a moment like that, but whatever it was, it was unforgettable.

Despite the fun Karena and I did have at the reunion, my sojourn in England was taking a toll on our relationship. As we feared from the start, extended periods of absence were not making our hearts grow fonder. We talked about it, but there wasn't much we could change. I wouldn't be back to stay for at least another few weeks, minimum, after which I didn't know where I'd go or what I'd do.

My lifestyle depended on how things went with the Starchild, and now its bone was being tested to see if I'd been duped by a hoax about its provenance. I had to prepare for the worst, which would be—at least in its initial stages, until I found new footing—a life of desperation that I wouldn't want to inflict on anyone, including Karena. I'd be broke and frightened and in no mood to share my anguish.

July 3 came with no word from the lab in Florida or Ken Pye. The wait dragged on for two more weeks. In the middle of July, I called Ken to ask him to call Florida to see if there might be a problem. He said we should give them a bit more leeway, which I grudgingly agreed to do.

Meanwhile, Andrea Cross called to say she was having

trouble convincing her faculty advisers to take the Starchild seriously. She asked if I'd consider bringing it up to her university to show it to her professors, and to speak to a general assembly of students that she'd coordinate. She hoped people there might learn to stop thinking the worst about the Starchild based solely on what they imagined it *had* to be (a fraud, a hoax, a deformity). She felt that if they saw it with their own eyes, as she did, a majority would arrive at a more open-minded view of what it might represent.

She said she'd arrange a place for me to stay if I'd cover my train fare north. I talked it over with Belinda, who considered it a good idea to expose it to serious—if skeptical—academics. On the previous few occasions when I tried to speak to university groups in the States, I found it impossible to get a tolerant hearing. Professors challenged by my "absurd" ideas would agitate graduate students to harass me, and they in turn would attend my lectures and heckle me to the point of total disruption. It wasn't an experience I looked forward to repeating, but Belinda paid the bills.

Two days before I was scheduled to leave, Andrea called in a state of agitation, telling me the president of her university called her into his office and shoved at her one of the flyers she'd distributed on campus to announce my lecture. He made it clear I would *not* be allowed to speak in the main hall she reserved, and my presentation would *not* be open to all students. He'd allow only a small lecture room to be used, and only "hard science" majors could attend. They would be the least likely to take the Starchild seriously.

"Bringing a thing like this on campus!" he fumed at her. "We'll be the laughing stock of every faculty in the country!"

I listened to her anger and frustration and embarrassment as she told me the story, and then she said, "I'll understand if you want to cancel now."

"Remember when we first met, I told you how amazed I was to find someone from a university environment with a mind open enough to want to determine the truth about something like the Starchild? This is what I was talking about. This is why I never try to speak at universities any more. Most of the minds inside them are closed against anything really new."

"But that's not the way it should be!" she blurted, becoming yet another victim of the never-ending war between youthful idealism and adult pragmatism. "Not at all!"

"Welcome to the real world, kid . . . welcome to my world."

"So tell me," she almost whispered, "does that mean you won't come? Because if you don't, I'll truly understand."

"I can't wait to rub their noses in it."

Bless her freethinking heart, all the restrictions did was raise Andrea's staunch Canadian hackles to full extension. She made sure the small room we could use was crammed with students, some standing along the walls. I was delighted with the turnout, and with my relatively gentle treatment at the hands of several skeptics who attended. Their questions after my lecture were all delivered in very proper British form, which was vastly more cordial than how it goes in the U.S.

The professors, too, were equally polite in their grilling, which was no different from questions I'd been fielding for years. By that point I was able to deflect virtually any hostility, overt or subtle, by simply focusing on the main points of disparity between the Starchild and a typical human.

In the end, I have to say, I think I won more than a few converts. At the very least, Andrea was ultimately allowed to write her paper about the Starchild, which was the point of the whole exercise from its beginning.

[Note: In Andrea's lengthy dissertation she outlined a wide range of human deformities that could account for the many physiological anomalies presented by the Starchild skull. She could not, however, offer a diagnosis to explain all of them occurring together, but stressed that is not uncommon with skeletal remains and congenital and developmental conditions. As for the fibers and the red residue found in the bone, they were beyond her purview and were not dealt with.]

Another useful development came through the husband of one of Grant's many social and business acquaintances. Dr. Matthew Brown was a dentist specializing in pediatric cases.

When Grant told Matthew's wife, Jennifer, about the Starchild skull, she told Matthew about it and he agreed to analyze the piece of maxilla I kept for just such occasions.

Dr. Brown X-rayed the maxilla piece to recover much more detail than was available in the original fluoroscopic X-rays.

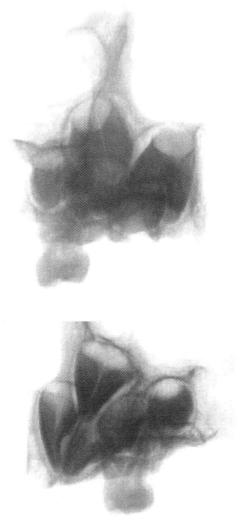

X-rays of the maxilla piece, taken in London in 2004. Notice it is missing the rear molar, which was removed at the BOLD Lab for DNA testing.

Applying his years of expertise and familiarity with juvenile teeth, he pronounced the maxilla as without question being from a child. That, of course, contradicts other specialists who believe the skull's robust cranial sutures and heavy tooth wear are more likely to be from an adult. I have to remain impartial, but I feel both sides are sincere.

Ken Pye finally called Florida in the last week of July. Sure enough, somehow our sample "fell through the cracks" when it came up for testing in the busy Fourth of July holiday week. That confusion led to it being slotted out of order. Their mistake was promptly corrected and sincere apologies rendered. To make amends, the test was done for half the usual cost.

When the results were in, Ken called to invite me to another meeting at his office outside London. This time Belinda drove me (I never dared to drive on the left side of the road—I had no doubt I'd end up in a collision), and we went as if to a hangman, fearing a judgment that we *had* been hoaxed.

Through June and July, I had struggled to figure out who would create such a hoax, and how it could be done. My first call was to Ray and Melanie Young, who naturally were shocked by Ken's suspicion. They were certain the couple who gave them the skulls would never be part of a hoax. Likewise, that couple was equally sure the original owners would never participate in a fraud. We couldn't imagine why or how such a hoax would—or could—be perpetrated.

I was torn by conflicting emotions. I had to respect Ken's reputation, and I did. But I also *knew* Ray and Melanie were solid, salt-of-the-earth folks who'd never knowingly be part of a hoax. I also tried to keep in mind the skull's many physical anomalies—still a deepening puzzle—and those bizarre fibers, which Ken admitted were an undeniable reality. I wanted to keep seeing a half-full glass, but I couldn't stop struggling with doubts. Eventually, even Grant's loopy idea about the Roswell aliens began to sound plausible.

When Ken called to set the meeting, I told him that Grant was no longer with me on the Starchild Project, and that my benefactor would take his place at the meeting. As Belinda and I settled into our seats across the conference table from Ken, his expression was more grimly serious than usual. He spent the bare minimum of time on polite introductory chit-chat with Belinda before opening a file.

Looking at me, he said, "Remember when I told you and Grant that Jean misinterpreted the the aluminum graphs because her field is organic rather than inorganic chemistry?"

I nodded. "Yes, I remember that."

"Well, mistakes work both ways, and now I'm guilty of doing the same thing. My specialty is inorganic chemistry, and I tripped over something organic. I owe you both an apology."

He slid two pages to us, which we read together. The Starchild's dating result was at the top of the first and was unambiguous: *900 years ago, plus or minus 40 years!*

A mountain of doubt lifted off my shoulders. *"Haaaaaa!"* I yelped, causing Belinda to flinch at my side. Such a very un-British surge of emotion caught her by surprise. "Right on the nose! The same as the adult!"

"Exactly," Ken agreed, "which is remarkable because you seldom find such strong synchronicity in Carbon 14's on separate samples. This is extremely convincing evidence to support the story that they died together as described."

I had just endured nearly two months of incessant self-recrimination for derailing my other interests in favor of what was shaping up to be a hoax. Now I was back in the game, breathing again, feeling the Earth move in a different way. I couldn't wait to call Ray and Melanie with the news.

"Listen, Ken," I said, "there are *no* hard feelings about this. We know you did your best; we know you gave us your best estimate. That's what we asked for."

"I do feel bad about wasting these past two months."

"They weren't wasted! Now we know the discovery story is almost certainly true, which has great value for us. And it tells us we're still in the hunt, which is terrific news! So, what now? Where do we go from here?"

Another grimace from him. "Well, it means we're now stuck with explaining the red residue, which has no plausible reason for being there. A number of useful tests can be arranged for it, and for the fibers too, but they all have to be paid for.

Will you be able to cover those expenses?"

He said that looking at Belinda, who replied: "That depends on what the tests are, what they can tell us about the skulls, and how much they cost. When I know that, I'll be in a better position to give you an answer."

"Fine," he said, "I'll work up a list of all tests with approximate costs. But understand beforehand that you have two problems. First, no tests I can recommend are as definitive as nuclear DNA. Critics can and will dismiss biochemical oddities as novelty quirks, freaks of nature. They'll say the same about DNA, but not as convincingly. DNA is everyone's yardstick, the thing we all take to the bank."

"And the second problem?" Belinda asked.

"No protocols exist for recovering either the fibers or the red residue. Both have never been seen before, so tools have to be created to extract them intact and uncontaminated. You realize, I hope, how microscopic they are."

"Well, surely *some* techniques apply here," Belinda said. "I mean, if scientists can capture a single sperm in a pipette and put it into a female's extracted egg, what *can't* they do?"

"You're right, of course, it can definitely be done," Ken assured us. "However, as is true with all the further testing, recovering those unique samples will carry a certain cost."

"Add it all together," Belinda said. "I need one lump sum to consider before I'll know if I can meet the obligation."

Ken couldn't tell that her stiff upper lip was talking.

Maldi-Tof test. Microtomography. Inductively coupled plasma spectrometry. Laser raman spectroscopy. Amino acid racimization and osteon analysis. Light element stable isotope analysis. Strontium and lead isotope ratios. X-ray powder refraction. Transmitted light (laser) microscopy. Electron probe and/or ion probe. And those were the *routine* tests!

Ken had found that no dependable costs could be fixed for testing the fibers and residue. Those would have to be created on the fly, so a blank check was required to even begin those procedures. That put them out of the question for the foreseeable future, but the routine ones were definitely in play.

The costs for those came to a minimum of 5,000 British

pounds, nearly 10,000 U.S. dollars. There was simply no way Belinda could cover those costs, not while maintaining her household and her other causes and charities.

"I'm sorry, Lloyd, you know how much I wanted to carry this case to a conclusion. But the simple truth is, I can't afford it. What I can do, for the moment, is keep you here in the house, with the hope you can find a way to bring in the needed funds from another source. That's our only option."

We'd already talked about this possibility, so she knew where I stood on it. "I've told you before, my mailing list doesn't respond well to my pleas for financial help. They've heard from Chicken Little too many times. Besides, you're better connected to rich people here in England than I am in the States. I hardly know anyone here—rich or poor!"

"Yes, but people I know think I'm a complete loon for how I spend my time and my money. What I've learned about most rich people is that as soon as they inherit, or as soon as a small pile begins to grow into a larger pile, they fixate on nothing more than making it grow larger and larger. Not to make the world a better place, to grow the pile."

"What about your siblings?" I knew they all were rich.

"They welcome me to family gatherings, but there it stops. I once tried to interest them in one of my causes, but it was *not* well received. Their attitude is, if I want to fritter my fortune away, it's not their concern; nor, for that matter, are my causes. My family is old money, and for the most part old money protects itself and grows itself larger or, at the least, stays put. I've ignored a kind of unwritten code. My siblings only donate to large 'proper' charities whose administrations gobble up most of the money donated, which is exactly what I battle against daily! It's infuriating, really; it drives me mad!"

To me it was a Monty Pythonesque dark comedy of skewed British manners. I never understood why Belinda chose unpopular causes that put her through such emotional and financial wringers. I knew why I chose the Starchild, but I didn't know why she did. I had benefited from her special bent of mind, her Quixotic spirit, but in all my months of living in her house I still didn't know a fraction of what made her tick. She was a blue-blooded enigma.

"If your own family won't help you, and they're all so convinced that you're a wacko, does it ever occur to you that . . . well, you might actually *be* a wacko?"

I said it smiling to let her know I was only kidding—sort of. But she took me seriously. "I see myself doing a job ordinary people would do if they could; if they knew it was a job needing doing. I'm a self-appointed representative of any person who isn't a scientist or expert of any kind, but who suspects there's more going on in the world than we're ever told. I like to think I represent those who don't habitually turn a blind eye to all of the lies and subterfuge committed by those in power. I believe ordinary people are like me, and I *hugely* resent being left out of the loop of actual, true reality."

"Okay," I said, "I'm with you so far."

"As for the Starchild, I represent those who want to know if a hybrid species might once have been on our planet at some point in the past. I feel I speak for that group when I absolutely *demand* to know about it, so I've put my time and energy and money into a person—you—who seemed to be as committed as I am to securing that answer. Now that I've lived in the same space with you for months, I know I was right. If anyone can do it, you can."

I couldn't understand why people kept saying that about me. Ray and Melanie . . . Karena . . . Belinda. . . . They all insisted I was doing as good a job as I possibly could, but it never felt like that to me. With everything continually dragging on and on, with nothing ever resolved, nothing ever final, I felt my record spoke for itself. I had to be judged a failure, and any number of other people would probably be handling this project better than I was. *But who?*

Who could I pass it off to before it dragged us all under?

CHAPTER THIRTEEN

NEW DIRECTION

Winds and waves are always on the side of the ablest navigators.
—*Edward Gibbon*

I had not asked my mailing list for money in well over a year, so they had gotten used to my silence. When I contacted them this time, I heard only a faint echo in response. We collected less than $2,000—not nearly enough.

Another thing I hadn't been doing was contacting media outlets. No magazine articles were appearing, no television appearances were arranged. In that sense I had placed the project on automatic pilot, especially after Ken Pye dropped his bombshell about the strong possibility of a hoax.

Feeling trapped and desperate, I tried to get on *The Richard and Judy Show*, the English equivalent of Oprah Winfrey's famed U.S. talk show. I had already been on with them once, when the skull was a novelty. Now that we knew about the fibers *and* the red residue, I felt we had more than enough to warrant a second visit for a more in-depth discussion. I thought it would be "easy-peasy," as the Brits liked to say.

I took the SEM photos and Ken's color photos to Cactus TV's studio to speak to the producer I dealt with before. He met with me for a half-hour, listening politely as I discussed the extremely unusual fibers and the reddish residue, and how that pair of anomalies made it highly likely the Starchild would eventually be proven to be not entirely human. It was just a matter of time, I assured him, before we had enough money and technical support to dot all the i's and cross all the t's to a level of scientific certainty.

"Sounds interesting," he said, all smiles. "Quite interesting, indeed. Don't worry, you'll hear from us in a few days."

After a few days, I did hear, and their answer was, "No."

Next I set my sights on newspapers in London, the home of tabloid presses. London has twenty-five dailies or weeklies, the dominant ones being the *Mail, Globe, Evening Standard, Guardian, Mirror,* and best of all, *The Times.* I assumed that competition among them, and with true tabloids like the *Sun* and *Express,* would create a willingness to push the envelope to be first to print a story that might prove to be of historic significance. I was certain at least *one* of London's twenty-five dailies would decide to cover the Starchild, and thereby publicize our need for further test money and scientific support.

Each paper had a *Science* section, so I tried to visit the editor of that department at the top ten papers. Naturally, some dismissed me outright, an American larking in England, wanting to talk about a highly unusual skull. However, a half-dozen were intrigued enough by my query to invite me to explain it to them in full. Collectively, the reactions at those meetings were best expressed by one editor who heard me out with his jaw hanging slack. When I finished and asked him for coverage in his paper, he spoke.

"What do the boffins think?"

"What are boffins?"

"Intellectual elite . . . experts."

"Ahhhh. Well, I can get some of your best mycologists to tell you it's worth a close look. I can get a topnotch forensic geologist. I can get a brilliant forensics student at—"

He raised his hand to interrupt. "I meant those like Richard Dawkins, who shape the public's opinion about science."

Richard Dawkins was a boffin from Oxford University who was widely-known in academic circles as "Darwin's Rottweiler," a wordplay on Charles Darwin's original rabid defender,

Thomas Henry Huxley, who immodestly called himself "Darwin's Bulldog." Dawkins was implacable when defending the ineffable correctness of Darwin's model of evolution. Anyone forced to tread more than lightly onto what he considered his home territory could count on being savaged by his greatly feared rapier wit and draconian personal style. The Rottweiler made the Bulldog seem like a neutered poodle.

"He'd probably piss himself laughing," I replied, "at the idea of a skull that may not be entirely human. You'd have to expect that because he doesn't know anything about it. He'd just be shooting from the lip, which is what all experts do when they don't know what they're talking about."

The science editor bristled. "Let me caution you, sir, that Richard Dawkins is a personal friend of mine."

He was also on TV regularly—with Richard and Judy.

Another blow came with the announcement in late 2004 of a discovery in the Philippines that was trumpeted by every kind of media all over the world. An 18,000 year old dwarf had been discovered on the island of Flores, and this species would be officially accepted as a new kind of human. Not without a fight, of course, because all new discoveries of that magnitude have to crawl to their rightful position over the cold, dead bodies of the entrenched status quo.

Sure enough, the bones in contention were soon "stolen" from their discoverers by a crusty member of the Philippine archeological mainstream, who promptly announced the discoverers had made a mistake and the beings in question were microcephalic dwarfs, not a new type of human. Does this sound familiar? It sure did to me.

Microcephaly is the opposite of hydrocephaly. The latter is expansion of the cranium and brain caused by internal pressure due to the improper draining of spinal fluids, or "water" (hydro), while the former is reduction (micro) of the cranium and brain due to genetic malfunction. In addition, both can be used as handy catch-all terms for dismissing any anomalous discovery that few hidebounds in the scientific community would want to deal with on its own terms.

As I learned about this incredible discovery, I watched,

fascinated, as experts excoriated each other about it, some yearning to free the genie of new knowledge from eons in its bottle, while others were desperate to force it back in, out of sight and out of mind from those not wanting to change long entrenched opinions, values, and prejudices. And the same held true with the Starchild, everywhere I took it.

Just imagine, for a moment, as I did on many occasions, what the world's reaction might have been if the Starchild's skeleton had been discovered on the island of Flores by credentialed anthropologists doing a certified, well-recorded dig. Better yet, imagine that the Starchild and the Hobbit were found lying side by side, in the same geological strata. What would be the reaction to such a mind-bending postulation?

Except for its bipedality, the Hobbit was clearly non-human. In fact, during a *60 Minutes* interview one of the Hobbit's discoverers stated that its bones were much closer to chimps in size and shape than they were to humans. Admittedly, the Starchild strays far from human in the opposite direction, but both are certainly equal in their degrees of strangeness.

Unfortunately, those not wanting to deal with the Hobbit (much less the Starchild!) simply announced that it *must* be the result of deformity because, as has been explained earlier, deformity is all-encompassing and so can be a repository for all manner of uncomfortable realities. Hobbit? *Deformity.* Starchild? *Deformity.* Like playing chess against a robot.

While my fruitless ordeal with London media played out, an email came in from a man named Kevin Morrison, who had read an article about me and the Starchild in a magazine from the U.S. He wanted to know more about it and asked some probing questions. His email address showed he was based in England, so I told him I lived in London and suggested that if he were in the city, or near, he could come to Highgate and see the skull copy for himself.

He wrote back to say he was interested, but he lived in Liverpool and would be stuck there with business for the foreseeable future. I wrote back to say I had a friend in Liverpool, Ron Jones, one of the most jovial, agreeable men I'd met in years. I'd visited him once up there, then passed by again on my way back south from Andrea Cross' university. So when I wrote that to Kevin Morrison, he replied that if I ever wanted to go back to visit Ron, on a weekend or at any time, he (Kevin) would pay for the ticket if I'd carve out enough time to show him the skull and have a chat.

I couldn't pass up an offer like that, so in another week I trained north to Liverpool to spend a relaxing weekend with Ron Jones while meeting Kevin Morrison. He turned out to be ex-military, like Ron, though Ron was ex-Royal Marine while Kevin was ex-paratrooper. Ron is short and solid, the kind of man you know would be useful in a street brawl if it ever came to that. Kevin is tall and imposing, with a soft voice, dead-level gaze, and hands the size of shovels. He's the kind thugs would never dare pick a fight with in the first place.

Kevin Morrison

Though very personable, Ron Jones is a die-hard skeptic, so he was firmly on the side of the Starchild being a deformity until undoubtedly proven otherwise. Fair enough. Kevin Morrison took the opposite approach. He thought, on balance, that the evidence pointed to it being a human-alien hybrid, and he'd assume that until proven otherwise.

I explained to Kevin the skull's entire history, including the many ups and downs I had been through getting it to that point. Then I told him the wheels were coming off the wagon in London, and I would be returning to the States in the near future if something didn't break our way soon, in a big way.

"How big do you mean?" he muttered, softly as always.

"Five thousand pounds minimum to initiate testing, but then I have to have enough to live on while it's being done."

"How much would you need for that?" he pressed.

"Ken Pye said it could take a year or more to work all the tests into his busy schedule, so to carry me that far I'd need in the range of ten thousand."

"Fifteen thousand total?" he said. "That would fund it?"

"Cheap at twice the price!" Which might have been true.

The meeting with Kevin was little different from a dozen I'd already arranged with people who, like Andrea Cross, learned I was in England and came to Highgate to see the Starchild material for themselves and talk it over with me in person. They'd invariably get excited and be struck by an urge to help me find people to fund it. However, when the rubber met the road, they quickly learned that people with money wouldn't put it into "alien" research.

What I couldn't know was that Kevin had not only the will but also the capacity to do something about it. When he told me he worked "in construction," and when I saw his size and demeanor, I assumed he was a hands-on builder. Wrong. He was a manager and a builder, which provided him a certain degree of wealth. He wasn't a rich man—Ron Jones seemed much better off—but, like Belinda McKenzie, Kevin Morrison felt a *desire* to contribute in a way almost nobody else did.

By then it was late October. I told him if he was serious he had to act soon because unless something major happened, December would be my last month in the house. Belinda and I had agreed to make a fresh start in 2005.

Kevin said he'd have an answer within a few weeks.

While waiting for things to develop in Liverpool, something happened in the house. Among Belinda's numerous pressing obligations, perhaps the most important, and the one she took most seriously, was the mental condition of her youngest child, Tizzy, in her mid-twenties and plagued with periodic bouts of schizophrenia. When severely afflicted, Tizzy had to be hospitalized. When feeling better, she came to live in High-gate. At that time she was in the house, and Belinda needed help with her and help with the house in general. A friend of the family, Tizzy's age, agreed to move in for three months as a full-time helper/companion/housekeeper/cook.

Her name was Amy Vickers, a well regarded second-year law student at Flinders University in Adelaide, Australia. She had always wanted to visit England, her father's birthplace, so on her "summer" break from school (winter in the northern hemisphere is blazing summer in the southern) she was coming to help Belinda with Tizzy, and to see some of England's winter highlights during her time off.

Amy Vickers shortly before her arrival in London in November, 2004.

It was my job to go to Heathrow Airport to escort Amy back to Highgate to make sure she arrived safe and sound. As any-one would after twenty-four hours in transit, she arrived tired and rumpled, but she was clearly the smiling girl in the photo she sent so I could recognize her: tall, very heavy-set, light brown hair, full-lips, and upturned gray (yikes!) eyes hinting at amusement or playful fun, or maybe mischief.

She defined "big and beautiful."

Subways were difficult to negotiate with luggage, so I arranged for us to take a bus into town. In that hour drive I found her to be amazingly self-possessed for a girl her age. I had been told she was a top student in her law class, and then her personal "class" became obvious. I couldn't help being smitten by her breezy style and crackling intellect, but at thirty-plus years her senior, and somewhat less than her physical size, I did nothing to indicate that.

Funny thing, though . . . she acted as if she might be a bit smitten, too. I chalked it up to her being exhausted and overjoyed—to the point of delirium—to be on the ground and close to her new home for the next three months.

Having completed her arduous journey, Amy Vickers took a brief nap and then, fighting jet lag, she had to prepare a meal for eight because one of Belinda's many focus groups had been invited to dinner. Without knowing the kitchen well, Amy went into it and two hours later blew all eight of us away with the quality of the meal she prepared. If I was smitten before, now I was gobsmacked. Her poise under such pressure was awesome, and her culinary skills stayed at the same astonishing level, day after day after day.

That old bromide about the way to a man's heart being through his stomach? I used to laugh at that, ridicule it, disparage it. Very quickly I became a fully reformed skeptic.

From that first bus ride, Amy and I simply clicked. Miracles happen. Some people win lotteries. We won this one.

Her room was my old room on the second floor, but she began to spend more and more time with me up in my top floor flat, chatting with me, getting to know me. We became friends, so I began to accompany her on jaunts to see the country. We went to Stonehenge, castles, plays. Our initial friendship slowly, steadily, solidly grew into much more.

She began reading my material, and in two weeks had burned through *EYKIW* and all of the articles posted on my website. In the house were a few videos of me giving presentations, which she watched and absorbed like the sponge she was. Much, if not most, of it was a revelation for her because she came from a conservative Christian background. She'd never heard of the Council of Nicea, or how the King James Version of the Bible actually came to be written.

For as brilliant as she was academically—there was no doubt about that—she was woefully uninformed about the world's realities, and the more she saw of my material, the more she understood and accepted that. Two questing minds were connecting in a way and at a pace they seldom do.

One day when I least expected it, Kevin Morrison called to say he had put together the money needed to cover the bone tests and most of my living expenses, but not necessarily all I'd need over the course of a year. I thanked him profusely and told him I'd see if Belinda might be able to help me fill any gaps. Kevin said he would be satisfied with an arrangement like that if we were, and of course we were.

Amy and I went to Liverpool to square it all away. It was her first outing with me as "the Starchild guy," as opposed to the man she knew in Belinda's house. She stepped right into the role of partner and held up her end of everything we needed to do. I couldn't have asked for a better teammate as we went over with Kevin the various details of our obligations for the money. Amy was a top student in contract law at her school, and her expertise was immediately and consistently evident. She was a godsend.

On the train ride back to London, she and I had our first serious chat about the future. It was the shank of November, a month since she had arrived. She kicked it off.

"What will you do now that you have the money you need?"

"Give it to Ken Pye and have him arrange whatever testing we can afford with that amount."

"No, I mean the living expense. Where will you stay?"

I shrugged. "With Belinda's tight money situation, I need to move out of the flat. She could rent it for a thousand pounds a month. Besides, it costs twice as much to live here in England as in the States. I'll probably just go back there."

"To do what? What will you do waiting for the results?"

This struck to the heart of something I'd been thinking about since Grant moved out and the handwriting on the wall became clear. "All along I've believed the stupidest thing I could do was commercialize the Starchild skull. But maybe six years is long enough to hold out. Maybe it *needs* commercialization to bring it to the attention of more people."

"What do you mean?"

I looked across the train seat at that incredibly lovely young face, so energetic and filled with the newness and stunning potential of adult life, so driven to seek out pure knowledge for its own sake. *Man, if I was only. . . !*

"I'm thinking of writing a book about these past six years. If Ken is right and the testing takes up to a year, there's no reason I shouldn't use the time to write a book. It'll give me something constructive to do, and it would give Belinda and Kevin a chance to recoup some of the money they're putting into me and the skull. It can be win-win all around."

She beamed the warmest, the most joyfully vibrant smile I felt I had ever witnessed. "I think that's a brilliant idea! You're a wonderful writer, and I can't wait to read it."

[Note: While many people contributed money to the Starchild Project, and those gifts were always useful and appreciated, the only three who provided signficant support were Belinda McKenzie and Kevin Morrison, and Pat Snuffer before them. However, several others were also notable contributors: Ron Jones, Kate Gooch, Andrew Johnson in England; L.A. Hotze in Germany; Karen Lyster in New Zealand; Debbie Payne in Australia; and in the U.S., Houston Anderson, Ken Elliott, and Tom & Pat Hungate were consistently helpful.]

When we returned to London, I shared the book idea with Belinda. She was all for it and promised to help in any way she could. I stressed to her, as I later did to Kevin, that their combined contributions would be paid back out of the book's profits, if there were any. We'd share equally until they were compensated. They agreed, and also agreed to make sure I had enough money to live modestly while writing it, which was fine with me. I had never lived any way but modestly.

Belinda drove me and Amy out to visit Ken Pye to deliver the money to get started on the inorganic testing he would be arranging and overseeing. I felt as light as a feather, as if the weight of the world was off my shoulders and into someone else's lap—someone much better equipped to handle it. Ken thanked us and promised he'd get those routine tests completed when and as his busy schedule permitted.

Toward the middle of December a minor crunch came. I'd been telling Belinda I'd leave before the holidays so I could be home with my family for Christmas. However, Amy wouldn't return home to Adelaide until the end of January, the length of her obligation to Belinda and to Tizzy. Naturally, I'd started looking for reasons to stay to be with her.

Christmas approached with me voicing no plans to leave, so I had to explain myself to Belinda, who was too proper to ask me outright but had to be wondering. It was no secret that Amy and I were spending all of our free time together, making it obvious something was happening between us.

On the anniversary of Amy's second month in London, I invited her to celebrate in a neighborhood French bistro. In two months we had become almost inseparable, and had also become immune to glances and whispers in public places. *Nooooo, she can't be with him! She must be his daughter!* Then she'd take my hand, or lean into me, or perform some other act of affection, large or small, and whoever was watching us had to deal with the truth as best they could.

I held up my wine glass to tip against hers. "To the best two months I've had since. . . . Well, since I can remember."

Her biggest smile blazed across the table. "To our *first* two months," she amended. "We've only just started."

Her suggestion of more to come was surprising. As a lawyer in training, she tended to choose her words carefully.

"Actually, this is something I want to talk to you about."

Her smile turned demure. "I bet I know what it is."

I gazed into her exotically upturned eyes, wondering how deep they actually saw . . . if they gave her a view into my aching-with-anticipation heart. "Take your best shot."

"You want me to consider staying with you and working with you instead of going back to Adelaide and law school."

That demonstrated one of the main reasons I already loved

her. Mentally she was always either stride for stride with me, or a step ahead—never behind. "Yes, that's exactly right. Is it a hopeless fantasy, or would you consider it?"

Her Aussie accent chimed. "I think about it all the time."

It was one of those seminal moments in life that comes along once in ten blue moons. "What's the verdict so far?"

Her smile switch turned up to dazzle. "I've fallen in love with you. Actually, I started falling in love on that bus ride."

Reeling with delight, I smiled as wide as my lips and face and emotions would permit. But then, as the import of her words sank in, I took a sip of wine as a frown crept across my extensive brow. "Have you noticed that you're young and beautiful, while I'm . . . well, neither of those?"

"You're old, Lloyd—old and gray and bald. But love doesn't let us reject things like that, does it? You have to accept the package it's wrapped in, like it or not. So believe me when I say this: I honestly love your entire package."

We gazed across the table at each other, already starting to work things out with our eyes. "You know," I said, "I have another package you might want to consider opening soon."

She lifted her wine glass, we tinkled their rims, then she parodied the best line in one of our favorite songs, *Walking in Memphis* by Marc Cohn. "*Man*, I will tonight!"

That was it for us. I had to tell Karena what had happened, that I'd fallen in love with someone else and intended to try to make a life with her. Due to the tenuous nature of our long-distance relationship, we were always keenly aware that we remained in play, subject to being swayed by the beat of a different drum. Still, that made it no easier when it happened.

Ending a relationship, no matter how prepared you both might think you are, is as emotionally painful as life experiences come. Ties to the past must be severed to allow room to embrace the future, and that process always hurts. But if the future you want to embrace looks as bright as mine did with Amy, you have to cut clean and work diligently to make the best of it. That was exactly what Amy and I intended to do, and we hoped Karena would graciously do the same.

Next, we told Belinda that Amy would join me in the flat and would also join me when I returned to the States, to be with me while I wrote the book. Belinda was predictably confounded, as would anyone faced with our huge age gap. However, hearing us explain ourselves—especially Amy's perspective—took much of the edge off her reservations.

"Lloyd is doing some of the most important work in the world right now," Amy explained, "but not nearly enough people know about him. He needs a much wider audience, and I want to do everything I can to help him get it."

"But. . . ." Belinda sputtered, "are you two fully certain you can . . . well, make a *go* of it?"

"Let me put it this way," Amy said, in her most matter-of-fact, lawyerly tone, "I'm not interested in starting a family any time soon, so I can afford to give him a few years."

Belinda nodded, then asked, "What about Australia?"

"I'll miss it sometimes," Amy admitted, "but this is more important. Besides, the world doesn't need another lawyer, and I don't need to go back to marry some insensitive dropkick who cares more about drinking with his mates than being with me. But Lloyd . . . he actually *listens* to me. I've never been with any other man who does that. He's exactly what I've always wanted, but I didn't know enough to realize it."

Belinda nodded. "He is good at listening, isn't he?"

Amy turned to me and flashed her most dazzling smile. "He's good at a *lot* of things, Belinda. I'm so very happy with him, and we want you to be happy for us."

Belinda hugged us and we cried a little, danced a little, drank a toast, laughed a lot, and knew in our hearts we'd never again be quite as close, or as filled with expectation for ourselves and the world's only Starchild skull. On that very special night in Highgate, London, we felt invincible.

CHAPTER FOURTEEN

THE FINAL HURDLE

If I have seen farther than other men, it is because I stood on the shoulders of giants.

—Sir Isaac Newton

Amy and I left for the States in late February, 2005, only a few days after my sixth-year anniversary since meeting the Starchild. Amy fell in love with New Orleans faster than she did with me, and gladly would have settled there if I were willing. However, in preceding years I had read articles in *Scientific American* and *National Geographic* about the extreme risks New Orleans always faced during hurricane seasons. I'd already endured five close calls with Category 3 storms striking nearby, and each new one that formed was a danger-drenched crapshoot. Disaster seemed inevitable.

I also knew Louisiana's state government was inept and corrupt in roughly equal measures (surely more than most states), which was matched by rampant cronyism in the New Orleans city government. Anybody paying attention to local gossip knew the U.S. Army Corps of Engineers did shoddy work constructing and maintaining levees and seawalls. Worst of all, state and local politicians siphoned off vast amounts of money allocated for Corps maintenance to fund local pork barrel projects such as golf courses and nursing homes. I had lived with that risk for six years, and now enough was enough. I just had a very bad feeling about what I knew.

My nephew's wife was in real estate on the famous "Emerald Coast" of Florida's panhandle, with its green water and snowy beaches of pulverized quartz rather than the usual granulated sandstone. We went to look at rentals and found one that seemed ideal for our purposes and was within our meager housing budget. It was a fully furnished one-bedroom apartment above a garage owned by Bob and Susan Mitchell, agreeable landlords who quickly became good friends. We were 200 yards from the Gulf of Mexico, but with enough general solitude to create the necessary writing environment.

Because our new place was furnished, I left most of my belongings in New Orleans in a storage locker. Included in it were five boxes of hanging folders stuffed with notes and other research material I intended to use when I got around to writing my magnum opus. Now that I would be writing about the Starchild Project, the opus was on hold, so I left everything there, taking out only some clothes and a few items that our new apartment didn't provide.

We moved into it during the first week of April. Eight hundred square feet, about the size of the flat in London. Home.

In early July, Hurricane Dennis bore down on the Florida panhandle, so we evacuated to north Louisiana to stay with my parents until the storm passed. My ill relatives had endured their afflictions through my extended stay in London, and now they were delighted to learn I was loved by, and deeply in love with, such a brilliant young woman whose cooking could easily compete with the finest in the world. She proved it to them in that layover from Dennis, and they ended up as smitten with her as I was.

In late August, the Gulf Coast was shattered by Hurricane Katrina. As the horror of it unfolded, I was led to believe that my storage locker had been inundated. Several weeks later, I learned that report was inaccurate and it didn't go under. My personal good fortune crumbled, though, under the weight of my feelings of profound regret for the people of New Orleans, who deserved so very, very much more from all of their elected and appointed officials at the state and federal levels.

For the rest of 2005 and into 2006, with Amy at my side, I worked at writing this book. While I did that, she went on a rigorous exercise and diet regimen that peeled off the pounds she had carried into London. By the end of our first year together she had dropped eighty pounds and was now flat-out gorgeous in the world's eyes, as she'd been in mine all along. Her discipline and commitment to the task of pushing herself into better shape were awesome to behold.

Amy Vickers in November, 2005 (165 lbs).

I finished in late February, 2006, one year after we arrived from London, and only days after the seven-year anniversary of my first laying eyes on the Starchild skull. I felt the manuscript was topnotch, probably the best I had ever written, making several complex scientific subjects accessible to non-scientists by using informal language and back-and-forth dialogue, both of which were standard in fiction but seldom seen in nonfiction. I felt the new manuscript was original, compelling, and easy to read. It should be well received by any number of mainstream publishers.

That done, it was time to secure an agent to represent it to those publishers, and of all the possible agents in New York, one immediately sprang to mind—Russell Galen.

In the field of alternative knowledge, the most famous, if not the most prolific, writer was Whitley Strieber. He wrote *Communion*, the mid-1980s blockbuster book whose stark cover of a Grey staring balefully at us embedded in our minds the iconic image of those otherworldly beings. Ironically, that image is far from the one abductees normally describe, including Strieber himself. However, the artist hired to create the cover produced such a riveting, commercially viable image, the publisher wisely went with it despite its wide divergence from how Greys are typically described.

Most abductees claim that while Greys have small bodies, they have large, bulbous heads with a noticeable crease at the upper rear, below which exceedingly small faces are made to appear even smaller by the bulk of the cranium. This is a perfect description of how the Starchild might well have looked in life—not the image on the *Communion* cover, which has a small crown and heavy lower face.

Comparison of "Communion" grey and artist's conception of SC image. (Both images used courtesy of Rob Roy Menzies.)

Recently, Budd Hopkins revealed that he and other serious researchers into the UFO abduction phenomenon use the *Communion* face as a litmus test for alleged abductees. If they show it to an abductee who then claims the beings they saw resemble it, their story is immediately suspect. However, if they reject the cover and begin to paint a word picture more like the Starchild image, then researchers know that in all likelihood they are hearing the truth.

Whitley Strieber's literary agent was Russell Galen, who had risen from early representation of Strieber and others to become one of the best agents in New York, certainly in the top ten. While he still represents anything written by Strieber and his wife, Anne, he represents no one else from "the fringe." Still, if you've written something you think is first-rate, you try for the best agent you can secure for it, so I contacted Russell Galen and asked him to consider it.

As soon as he heard the words "human-alien hybrid," he was ready to end our exchange, but I convinced him to take the time to check the website. I also assume he checked with Whitley Strieber, whose name I dropped because Whitley had seen and held the Starchild in his own hands soon after I began trying to figure out what it was, back in 1999. Whitley *knew* it was a real, true bone skull.

Eventually, Russ agreed to take a look at the manuscript when he could find the time, which I considered a huge step in the right direction. However, several weeks later he informed me that he could not represent it. He was sure no mainstream publisher would touch it, or, if they did, they'd take no more than a passing swipe at it. They would never take it seriously. He wouldn't tell me why he was so certain of that, but he sounded convinced.

This was yet another crushing blow to my own spirit, but now I had Amy's feelings to worry about. In late April, while Russ was evaluating and our hopes were as high as they could be, we got married. We did it knowing full well the universal reaction to a couple marrying with an age spread like ours, but Amy had become such an integral part of my life and work, I really had no choice. It was marry her and make her an official partner in what I was doing, or risk her returning to Australia to resume law school. By then life without her was unthinkable, so we crossed our fingers, said our vows with a video linkup to her family in Oz, and took the plunge.

Amy Vickers and Lloyd Pye on their wedding day.

Russ Galen's rejection was a bitter pill to swallow, but it propelled me toward another top New York agent, Jane Dystel, who had established herself near the top of the heap when I began writing thirty years earlier. Her father, Oscar Dystel, had more or less invented the rack paperback as we know it, so in publishing circles they both were icons. Gaining her interest was a great coup that relieved the stress Russ caused.

Jane invited me to send a manuscript in early June, with the usual promise to get back to me about it within a few weeks. A few days later, I was called by a producer from the *National Geographic* cable TV channel (NG), asking if I would agree to be filmed with the Starchild skull for a show they produced called *Is It Real?* The NG channel wasn't carried where we lived in Florida, so while I knew it existed, I had never watched it. However, I remembered from my youth that anything with *National Geographic* on it—whether book or magazine or film—was absolutely first class.

They explained that they were planning an episode of *Is It Real?* based on the "Ancient Astronaut" theory first espoused by Erich von Daniken in the late 1960s in his world-famous bestseller, *Chariots of the Gods?* I asked why they were calling me, and they said alternative knowledge insiders had told them the Starchild skull was the most extensively tested relic or artifact ever put forth as possibly having an extraterrestrial origin. That being the case, they felt their show would not be complete without presenting it.

My earlier experiences with *Extra* and *The Learning Channel* had turned me off of TV documentaries in general, so I told the *NG* producers I wanted some time to think it over. Then I emailed several friends in the field, asking for input. Soon I was inundated with messages about how bad the *Is It Real?* series was in terms of giving its guests a fair shot to express their views without being contradicted or ridiculed by the ubiquitous mainstream "rent-an-expert."

Everyone insisted that the *National Geographic* name and reputation meant nothing relative to these shows. When it came to their new cable channel arm, they were treating it as a business, no different from those at the *Learning Channel,* the *Discovery Channel* and even the *History Channel.* None of those—nor any other domain of mainstream media—could afford, could dare, to take me or the Starchild seriously. They were too cowed by scientists and religionists, both of whom would erupt in vociferous outrage when any of their sacred cows were gored on national TV. There was simply no way our side could get a fair and open hearing from such people.

I contacted the *NG* staff and relayed to them what I'd been told about their show, and to tell them I wouldn't participate. They were flabbergasted, I'm sure, because everyone in the field of alternative knowledge understands the deal you have to make with the Devil to get your opinions aired on cable channels, much less on network TV. You have to sell out to the system to get even a portion of your message heard, knowing full well you'll be hammered unfairly—even maliciously—through the course of delivering it.

The *National Geographic* producers went into high gear, promising they would definitely take the Starchild seriously and give it a fair and impartial hearing. Sure, they admitted, contrary experts would have to be consulted, and they would have to be allowed to speak against it. However, that was only because fairness required them to always present opposing views to give their shows "balance."

Balance, I pointed out, was an extremely flexible term that could be slanted sharply or gently against any position at all. They promised they would slant as gently as possible against me and the Starchild, then they asked me to give it some thought and please take their promises to heart.

In discussing the show's parameters as they envisioned it, they said they wanted to interview me, the skull's owners, Ray and Melanie Young, and then at least one scientist among those who had dealt with it. I suggested Jason Eshleman at Trace Genetics, and they agreed he'd be perfect.

In the three years since Jason and Ripan tried so hard to sequence the Starchild's nuclear DNA, we hadn't communicated very often. There was no point in doing so until a new breakthrough was made in DNA recovery. But this, I knew, was a special situation that warranted a phone call from me to see if Jason would be willing to be filmed and identified to a wide audience as the geneticist who did the DNA testing of the Starchild skull's bone.

Jason did not lack for nerve, so he said he'd be happy to explain what he'd found during the many attempts he made at sequencing the Starchild's nuclear DNA. He still had to say he believed degradation was the problem behind his lack of recovery, but I knew that going in. Then, toward the end of our conversation, he casually dropped a bombshell on me.

"By the way, have you heard about the new sequencing technique being applied to the Neanderthal genome?"

"No, should I?"

"It's been discussed in some technical journals. A place called 454 Life Sciences in Connecticut has found a way to sequence DNA in a base-pair by base-pair arrangement. It's being applied now to a Neanderthal's DNA. They're already closing in on the first million of its three billion base pairs."

I felt faint. This was exactly what he and Ripan had discussed with me and Karena three years earlier, saying then they thought it would be three or four years before a breakthrough of any magnitude was made. Now here we were, three years exactly, and the big breakthrough was a reality!

Holy Maloley!

"You're kidding! They've done that! They've done away with primers? Why didn't you guys *tell* me? Ripan promised he'd call as soon as something like this was in the wind!"

"Didn't you know? Ripan has moved on from here. Amicably, I should add. Some personal things came up for him, so he needed a change of scenery. I miss his company and his help."

I barely heard Jason's words, reeling with the possibilities this opened up for the Starchild Project. Now we could legitimately anticipate a final, yes-or-no end to the mystery at some point in the future, and when we did, it would be conclusive. Once we had every base pair in hand, we could tell *exactly* how far from normal—or how close—the Starchild was. That determination would be the end of the line for what I had to do with it, but if we got the answer we now expected, nothing would ever be the same again for anyone on earth.

In the sciences, especially, many senior members would simply retire rather than endure the indignity of facing up to their errors and graciously accepting the new reality. And who could blame them, really? Just let the young professors, technicians, and graduate students reinvent themselves while they were still able to do so. Once the ossification of dogma has set in deeply enough, the impact of something *that* new would shatter them into tiny fragments. Yes, leave it to the few who would leap from astonishment at it, and to the ones with a passion for historic events and tumultuous times.

But when might those times start?

"Please give Ripan my very best wishes when you talk to him next. Now, tell me, how do you see this impacting the Starchild? What can we expect, time wise?"

"I'm afraid you're still in for more of your long wait. Right now, everybody's effort is being directed at the Neanderthal genome, and that won't be complete for two more years, until the end of 2008. Next in line after that will probably be Otzi or the Hobbits. That race is too close to call right now."

I knew either one would have a lot of scientific support.

"After those two get underway, though," Jason went on, "the field opens up quite a bit. If, between now and then, you can make enough people aware of the Starchild's case, maybe they can combine to create enough pressure, enough demand, to warrant utilizing the new technology."

"Oh, geeze, that would be *perfect!* What will it cost?"

"By then? Who knows? Maybe $200,000—maybe half that. It depends on how efficient they get at doing the recovery and sequencing, which determines the amount of time needed to complete an entire genome."

I could understand that. Three billion base pairs was an intimidating number in anybody's league. "When our turn comes, whenever that is, what time frame do you see for carrying it out? Can you give me even a ballpark estimate?"

"Well, let's say it takes two years to do the Neanderthal. In that time, dozens of economies of scale and technique will be found to make the process work much smoother and go much faster. That's how it was with the human genome project, and I'm sure that's how it will be with this, too."

"So?"

"So I think that if everything goes right and you're able to get the Starchild's case front and center in the public eye, we could start on it by 2009 or 2010 at the latest. And we might be able to have an answer within a year, maybe less. The learning curve on these things is remarkably steep."

"By the end of the decade, then?"

"If everything goes as it should, I think you can count on that as a solid time frame. Probably by the end of 2010."

Every inch of me was trembling from the excitement of it. "Jason, I can't tell you how incredible this news is to me. It changes everything! Now I can come out of the woodwork and start letting people know what we've been sitting on."

"Wasn't that what you were doing with the book you were writing? I mean, publishing a book is coming out of the woodwork in a pretty splashy way, isn't it?"

"To some degree, yes, but it had no bottom line, no finality. Now we have a finish line looming in 2010, and that's a whole different ball game. This is something people can get energized about, excited about, debate pros and cons—even take bets on! I'm sure once the testing is underway, you'll see a betting line on it in Las Vegas. It'll be like a Super Bowl!"

Maybe I was a bit delirious with the possibilities, but how could I not get cranked up over this news? A few minutes earlier I was stuck in a dark tunnel of uncertainty with no end in sight. Now the end, though still some distance away, was at long last visible. It was Christmas in June.

"Will you do the *Is It Real?* show now?" Jason asked.

"Sure, why not? I have to start getting the word out, and this is as good a place as any to start. The book will take a lot longer anyway. A show like this films and it's on the screen six months later. You sell a book and it comes out a year later if you're lucky; sometimes it's a year and a half."

"You'll have a three-year cushion," he reminded me.

"Yeah, but it's a *big* world out there. If I'm going to motivate enough people to put enough pressure on enough scientists to bump the Starchild ahead of their pet projects to use this new technology, I have to get busy!"

"Good luck," he said. "I'm pretty sure you'll need it."

In late June, Amy and I flew to El Paso to film our part in the "Ancient Astronaut" episode of *Is It Real?* The field producer was Nancy Donnelly and her assistant was Owen Palmquist. I liked them both and we had a good day filming in Ray and Melanie's house, and in the environs of El Paso. I can't fault anything they did that day. Everything was first class, and I feel they bent over backward to make sure our filming experience was as good as it could be. Everything seemed on a solid upward trajectory with this show.

Amazingly, on the day we arrived for the shooting, Jane Dystel contacted me to tell me she would handle the book. That was outstanding news, exhilarating news, because of her sterling reputation. However, she did make a couple of comments in a later conversation that gave me pause.

At one point she said this would be her first book of this

type, so it would be something of a learning experience for her. And she also said, "This book is the kind that either disappears with barely a ripple, or it becomes a huge bestseller worldwide. I don't see a middle ground for it."

"Can we go on the assumption it will be a bestseller?"

"I wouldn't take it if I didn't think it had a real chance. You're a good writer; that usually makes all the difference."

Something did make a difference, but not my writing.

A great deal had to be done to prepare the book's manuscript for submission. Nonfiction books are usually sold on the basis of an outline known as a "proposal." Proposals are popular with editors because they allow buying decisions to be made without having to read much. Nowadays, the "artistic" aspects that were once a part of publishing are gone. Today it is a conglomerate-dominated business based on a philosophy of "profit at all costs." Thus, no one at the highest levels of the editorial game has much time to read, so that time is saved for novels, which are harder to reduce to outlines. Nonfiction is best presented as a proposal.

Jane Dystel's original plan was to submit the Starchild book as a proposal, even though it was written. I set about creating an overview, a chapter-by-chapter outline, an author's biography, a table of contents, and, with Amy's help, a marketing plan. These required weeks to get right. Then Jane decided to leave off the outline part and submit the entire manuscript in its place, in case an editor was motivated enough by the overview and the marketing plan to delve right into it. In a non-calculated irony, we ended up with a hybrid submission for a book about a possible alien-human hybrid.

Twenty submissions went out in September, and in a few weeks we knew the worst was happening. Of those twenty, only two resulted in offers and both were insultingly low, making it clear they had no faith in the book's chances in the marketplace, or in my ability to promote it. In fact, I doubt if any of those twenty editors actually read the overview, much less the manuscript. I don't think anybody read anything. I think they took one look at the subject, couldn't imagine a promotional pigeonhole for it, and rejected it outright.

I knew an unusual book like this would be a problem. Where does it fit on bookstore shelves? In New Age? No. In Paranormal? No. In Science Fiction? Definitely not! In pure Science? That was closest, but it wasn't footnoted or referenced or overflowing with jargon. It was a science-based book that would actually be fun to read, so where would *that* fit?

Amy and I felt the pressure of having more than a year of strenuous effort swirling down some kind of unseen but grimly efficient drain. It wasn't much of a honeymoon.

As that played out, Nancy Donnelly at *National Geographic* notified me that November 27 had been chosen for the debut date of the "Ancient Astronaut" episode on *Is It Real?* I got with Jane Dystel and we formulated a strategy. We'd list editors who hadn't already rejected the manuscript, then inform them about the upcoming show to see if they would watch it as a prelude to requesting a proposal submission. After viewing the show, if they were intrigued by the Starchild's story, they could let Jane know and she'd forward it. Several editors agreed to go along with that plan.

Unfortunately for us, an extreme element of risk was involved. It required that the show be as respectful and balanced as Nancy and Owen had promised it would be. If it wasn't—if it was as hokey as every other episode in the *Is It Real?* series—that would create an insurmountable obstacle to selling the book to anyone in New York for anything like what it might be worth. So we crossed our fingers and toes, pulled on our lucky socks, and waited for November 27.

From my perspective, the show was a disaster. Giorgio Tsoukalos, Chairman of the A.A.S.R.A. (Center for Ancient Astronaut Research) and the publisher of *Legendary Times*, was another guest on the program. He had spent part of his Thanksgiving holiday watching every other episode in the *Is It Real?* series, and he assured me that we were treated with more respect than anyone else from our field in any other episode. With that said, Giorgio produced a 5,000 word document chronicling all that was wrong—technically and otherwise—with the show. I might have done the same if he hadn't done such a cogent, and thorough, job of it.

Watch it if you must, but be prepared for nothing out of the ordinary for this type of show. There is a format they have to keep to, and I believe in my heart of hearts that a good person like Nancy Donnelly did all she possibly could to mitigate the debunking job that she, as its producer, had to do on us. I have no hard feelings against her, none whatsoever.

The bottom line, though, is that the show killed us with the editors. After it played, not one of them contacted Jane to ask to see the book, and I can't say I blamed any of them. I'm sure I would have thought I was a humbug, too. So in regard to the book manuscript, there was nothing left to do but call in the dogs and water down the fire.

There would be no sale for it in New York.

In early December, Amy and I sat down after the latest in the series of defeats we had suffered in 2006. Try to imagine your first year of married life going as badly as ours had.

"What can we do now?" she moaned, her gray eyes brimming with tears. "We've worked sooooo hard, but now everything is just blowing up in our faces. I know you've had years of living like this, but I'm really, *really* not used to it and it's driving me crazy. Can't we make it stop?"

"Honey, we *can't* make it stop. We have to go on living."

"Not like this! I hate this constant waiting for other people to make decisions that determine what we can and can't do. I want us to take it all into our own hands. We're smart, we're capable, we both work hard. Why don't we just publish the book ourselves, like you did before?"

She was right, I did self-publish *Everything You Know Is Wrong*, which is why I'd been determined to find a publisher to take on all the onerous jobs and grunt work that go with self-publication. I'd never worked harder than when I published *EYKIW*, and I wanted to stay as far away from that grind as I could. But she was making a compelling point.

"I've told you how difficult that was," I reminded her. "It became this *thing* that consumed every minute of my time."

"Yes, but that was when you were by yourself, doing everything alone as a publisher. That's too much for one person to handle. But now you have a partner, and if we work together, we could make a business out of it! We could become a small publisher and put out all your other work in addition to this one book. We could create something we can nurture and grow, instead of a one-off and hope for the best. We could bring out *all* of your books—old and new alike—in the best way for them and for you, not in the way the suits at some big corporation think might be good for their bottom line."

She paused to hold my eyes and make sure I appreciated her seriousness. "We can *make* this happen if we decide to."

I gazed at her beautiful young face resting atop the lean body she had chiseled out over the past two years. The agony she put herself through to gain that had shown me she had the heart of a warrior. If I had to go back into the relentless grind of self-publishing—the never-ending responsibilities of promoting and selling—it needed to be with someone who had her kind of mental discipline and emotional toughness. And now, after these two years with her, I knew I could count on her to give 100 percent effort, all day, every day, for as long as necessary to achieve any goal she set for herself.

"It'll make law school seem like a vacation," I cautioned.

"I don't care! I'm tired of waiting for other people to decide our fate. If we have to go down, let's go down fighting, not sitting and fretting. Let's make the decision and *do it!*"

This book in this form is the result of her prodding, and now it's up to you, its readers, to determine its fate. As the old saying goes, if you like it, tell your friends, and if you don't, tell us. We're always willing to try to make it better.

And the "Bell Lap Books" you see all over this book? That's Amy. Bell Lap Books is her company. And in case you're wondering, a bell lap is the last lap in any long, difficult race.

We're very much hoping that this book's publication is the beginning of the Starchild skull's last—and best—lap.

EPILOGUE

All truth passes through three stages: first, it is ridiculed; next, it is violently attacked; finally, it is held to be self-evident.

—*Frederick Schopenhauer*

It's now more than eight years since I first saw the Starchild skull in that Holiday Inn lobby in El Paso. I've had a long, strange, enlightening, but mostly frustrating ride with it. Now all my hopes are pinned on this book's ability to reach a wide audience and thereby bring the skull to the attention of scientists who, if they would muster the courage and conviction to act, could mount an Otzi-like campaign of discovery to learn once and for all who or what the Starchild was, and where it might have come from.

What more is required? What more could we do, or should we have done? It's a genuine bone skull with the general shape of a human, but the bone is half as thick as it should be, weighs half as much, is filled with extra collagen, is more durable, is riddled with unknown fibers, and teems with strange red residue. Any one of those could be dismissed as an anomaly, but together they add up to something worth serious, detailed, costly analysis. And that's just the bone!

Its shape is equally anomalous. Reconfigured eye sockets, no brow ridge, no frontal sinuses, no inion (external occipital protuberance), a reconfigured and repositioned neck, great-ly reduced lower face, and a greatly expanded brain. Again, each can be dismissed as one-off freaks of nature, but taken altogether and coordinated so exquisitely—that warrants a mobilization of resources to learn beyond doubt what makes it so unusual in so many aspects of its physiology.

Courage is indeed what is needed, from every aspect of authority in our culture—science, religion, government, corporations, and the wealthy. All of society's leaders are obligated to facilitate what needs to be done. However, before they can take even the first step, they have to confront and deal with the frightening notion that beings with higher intelligence than ours might not merely exist "out there" in the vastness of the universe, but exist close enough to have visited and left a calling card of sorts in a dusty desert 900 years ago.

Despite the Starchild Project's widespread lack of support prior to publication of this book, its future is promising. Its biochemistry was indeed explored under Dr. Ken Pye's stewardship, and as expected, the results were always interesting and occasionally unprecedented. However, as Ken predicted, it was never enough to prove—as a successful DNA recovery would *prove*—that the skull was anything other than a highly unusual, one-off freak of nature. It certainly wasn't enough to persuade the Starchild's critics to delve into it with open minds. Those minds have to be pried open with crowbars.

[Note: Ken Pye's summary is provided in Appendix III.]

For the Starchild, and for me, it has always been about nuclear DNA, the only thing that can give indisputable proof of the skull's true genetic parentage. Toward that end, we rejoiced at the news of the revolutionary sequencing technique developed by 454 Life Sciences, which permits total recovery of an entire genome in a base-pair by base-pair array. This means that eventually—whether in two years or three years or even five years—*eventually* the Starchild's entire chromosomal package can and will be recovered, with all 25,000 genes from its mother *and* its father laid out in full, to be compared side by side, in every detail, with normal human mothers and fathers. That comparison is what we've been working for.

If—and be sure to understand that this will remain an "if" until the 454 technology is applied to the Starchild's DNA—if such a side by side comparison can be made, geneticists will be able to find exactly which parts correspond to the human genome, along with those that deviate from it, and especially the degree of deviation relative to humans, higher primates and, yes, by then even Neanderthals.

Those differences from humans are the critical discoveries, for Neanderthals *and* the Starchild. Where on the evolutionary line between humans and chimps will Neanderthals find their place? The range of difference between humans and chimps is now set at between one and three percent of their 3.0 billion base pair genomes. That means 30 to 90 million base bairs of difference, most of it in so-called "junk" DNA. Still, junk or not, that's an awful lot of base pairs to be able to manipulate in a genetics lab if you know what you're doing.

My personal feeling is that the Neanderthals will be much closer to chimps than humans. I don't think Neanderthals will turn out to be our second, third, or even fourth cousins, much less the literal kissing cousins many anthropologists insist they had to be. I think that is extremely unlikely.

As for the Starchild, will its difference be .001 percent? That would be quite close to human. Will it be .5 percent? A significant difference, to be sure. Or maybe 1.0 percent? That would cause people to sit up and taken notice. 2.0 percent? Now we're talking something radically different from human. 3.0 percent? That's as different from humans as chimps are. Would that qualify as at least partially alien? It would in most ledgers. And 5.0 percent? 10 percent? 15 percent?

Someone has to draw a genetic line in the sand, beyond which if the Starchild crosses, it must be considered an alien, nothing more or less, plain and simple. Many people will do all they can to avoid that, but the numbers won't lie. They'll say what they'll say, and we'll all just have to cope with it.

That day of final reckoning could well be one of the most important in history—or a hugely embarrassing whiff for our side. To echo one of America's most famous wits, Yogi Berra: This won't be over until it's officially over, and that, according to Jason Eshleman, won't be until sometime around 2010.

We've already endured eight years of this interminable wait, so we can definitely hold out for as long as we have to. All we need is—here comes that word again—the *courage* to reach our highest potential. And, as with climbing any mountain, even the most fearsome on the planet, or the most fearsome we can imagine, it all begins with the *intention* to climb, and the *desire* to succeed, then putting one foot in front of the other until there is nowhere left to go.

Once again, recall Rudyard Kipling's stirring poem, "If."

> *If you can trust yourself when all men doubt you,*
> *But make allowance for their doubting too;*
> *If you can wait and not be tired by waiting,*
> *Or being lied about, don't deal in lies.*
> *If you can hear the truth you've spoken*
> *Twisted by knaves to make a trap for fools,*
> *Or watch the things you gave your life to, broken,*
> *And stoop and build 'em up again with worn-out tools.*

If it was easy, anyone could do it; if it was easier, the right people might even try. But it isn't and they won't, so if the final chapter of the Starchild's remarkable story is to be written, *you*, the readers of this book, must make it happen. Word of mouth is a transcendent weapon, and that is what *you* must supply to friends and acquaintances who have a known, or a perhaps latent, interest in such controversial material.

Remember Bertrand Russell's trenchant quote: *What men want is not knowledge, but certainty.* I hope and believe the Starchild's story has sufficiently demonstrated that fundamental truth. I also hope and believe it has demonstrated that, occasionally, often when we least expect it, our emotions are stimulated enough to allow us to stretch beyond our safe certainties to seek new and distant horizons.

No horizon is more distant, or less certain, than confirmed life from beyond Earth, yet it now lies within the reach—and perhaps the relatively easy reach—of us all.

Π   Π   Π

If you'd like to join the Starchild Project in attempting to explore—and perhaps complete—that awesome, mind-boggling reach across reality as we know it, please add your email address to the website's mailing list: www.starchildproject.com. You will receive periodic updates about whatever is happening with or to the Starchild skull. That will put you on the cutting edge of events as they unfold, where you can stay until we all find out how this incredible story ends. See you there!

APPENDIX I

Preliminary Analysis of a Highly Unusual Human-Like Skull.
Dr. Ted J. Robinson, MD, LMCC, FRCS

The skull in question has a provenance that is not verified at present. That situation might change in time, but for now all that can be said with certainty is that the skull is real, it is comprised of calcium hydroxyapatite (the essence of all mammalian bone), its parts are configured naturally (not cobbled together or in any other way hoaxed), and it presents numerous physical anomalies that do not conform to standard norms for human skulls.

The skull remained in my possession in Vancouver, B.C., for the better part of one year. I was given complete discretion to study it in any way I saw fit. My analysis derives from extensive examination of the skull itself, combined with detailed analysis of X-rays and CAT scans. I have shared these data with colleagues who have given opinions that will be mentioned in this document as their input becomes relevant.

In general, the skull has the basic components of a human skull: i.e., a frontal bone, two sphenoids, two temporals, two parietals and an occipital. These bones are, however, markedly reconfigured from the shapes and positions such bones usually have. In addition, the bone has been reconstituted to an equally marked degree, being half as thick as normal human bone, with a corresponding weight that is half of normal. The reconfigurations and the reconstitution are uniform throughout all axes and in all planes of the skull. There is no asymmetrical warping or irregular thinning that is the hallmark of typical human deformity.

This skull's morphology is so highly unusual as to be unique in my forty years of experience as a medical doctor specializing in plastic and reconstructive surgery of the human cranium. Because of its uniqueness, I undertook an extensive review of literature on craniofacial abnormalities, which failed to uncover a single similar example. In short, it seems to be not only unique in my personal experience, but also unique through the history of worldwide studies of craniofacial abnormalities. This is significant.

Specialists who examined the skull, X-rays, and CAT scans were:

Dr. David Hodges, Radiologist, Royal Columbian Hospital, New Westminster, BC

Dr. John Bachynsky, Radiologist, New Westminster, BC

Dr. Ken Poskitt, Pediatric Neuroradiologist, Vancouver Children's Hospital

Dr. Ian Jackson, (formerly of Mayo Clinic), Craniofacial Plastic Surgeon, Michigan

Dr. John McNicoll, Craniofacial Plastic Surgeon, Seattle

Dr. Mike Kaburda, Oral Surgeon, New Westminster, BC

Dr. Tony Townsend, Ophthalmologist, Vancouver

Dr. Hugh Parsons, Ophthalmologist, Vancouver

Dr. David Sweet, Forensic Odontologist, Vancouver

Dr. David Hodges, a radiologist, stated that none of the skull's sutures were fused at the time of death. Dr. David Sweet, an internationally renowned forensic pathologist at the University of British Columbia, was of the opinion that the skull was of a 5- to 6-year-old, based on dentition in the right maxillary fragment. Other specialists who examined the skull have disagreed, but I support Dr. Sweet in his belief that this was the skull of a 5- to 6-year-old child.

Dr. Bachynsky saw no evidence of erosion of the skull's inner table. Such erosion would be consistent with a diagnosis of hydrocephaly, so this condition can be ruled out as a cause of the abnormalities expressed. Hydrocephaly also causes widening of the sutures, again not expressed here.

There was consensus agreement to both of these observations by other experts conversant with these features.

Dr. Kaburda carried out special three-dimensional X-rays which measure thirty fixed points in any skull, allowing for comparison of any particular skull to the established norm. These accumulated results were compared to a statistical analysis of 100 human skulls. This skull was often found to be more than ten (10) standard deviations outside the norm, i.e., the statistical center of a bell curve. This is another strong indication that the skull in question is unlike anything previously seen or investigated.

Doctors Townsend and Parsons examined the orbital cavities and concluded that the being may well have been sighted. If so, its visual structures deviated strongly from the norm. The cavities, while astonishingly symmetrical, were less than 50 percent normal depth. The optic foramen, which carries the optic nerve from the brain through the orbital bone to the eye, is nearly an inch lower than in a normal human skull. In the normal human orbit, the superior and inferior obital fissures connect to very consistently form an L-shaped cleft through which blood vessels and nerves enter and exit the orbit. In the Starchild skull, the cleft is a straight line.

Attachment points for the muscles that control an eyeball's movements were still to be felt on the inner surface of the orbit, indicating that a ball rather than some other mechanism was its most likely expression. If indeed the sockets held eyeballs, those of normal size would greatly protrude from the face, creating a serious liability of damage during routine activity. Because the eyeballs occupy a position lower in the face than is normal, and they rest in a socket markedly reduced in rectilinear shape and depth, they should have been significantly reduced in size. In either case, however, large eyeballs or small, they would require upper lids three or four times more extensive than normal upper lids to be lubricated in a manner necessary for human eyeballs to function properly.

Doctors Hodges and Poskitt found the brain inside the skull to be abnormally large. This was determined by lining the intracranial cavity with a plastic bag then filled with Niger birdseed. This gave a size of 1600 cubic centimetres, which is 200cc larger than the typical adult size of 1400cc This is even more unusual because the skull's size compares most favourably with a small adult or a child of about 12 years old. This extra brain capacity is apparently due to deep shallowing of the eye sockets, a total lack of frontal sinuses (not even vestigial bumps are seen), and significant bossing (expansion) of the upper rear of both parietals.

Doctors Hodges and Poskitt also observed that the extreme slant of the rear parietals and the occipital bone challenges whether this skull could have contained typical brain matter, and casts further doubt that its cerebellum was typical. In a normal skull, the cerebellum rests at the base of the cerebrum, supported by the internal occipital protuberance and the twin flares of the sagittal sulcus and transverse sulcus.

With this support mechanism, over the course of a lifetime the cerebrum's weight does not press down onto the cerebellum and distend it such that it will cease to function properly. In this unique skull, however, the entire weight of the brain slants directly down onto the area that should hold its cerebellum. Instead of the rounded area typically present for support, there is a wedge-shaped area perhaps one-quarter of normal. Furthermore, the internal protuberance and sulcus ridges are reduced. What effect would the weight of a notably amplified brain have on an unsupported cerebellum carried at least into childhood? It presents a genuine conundrum.

Personally, I was most concerned with determining how the rear of the skull could have become so flattened, from the atypical fossa (depression) in the sagittal suture between the parietals, down to the foramen magnum opening. This could not have been caused by any kind of flattening or binding device because the surface of the occipital reveals the subtle convolutions inevitably present in unaltered skulls. Skulls that undergo any kind of shaping technique will always reveal such technique with a distortion of the bone surface. Lacking even a hint of evidence of shaping, and of the unnatural or premature fusing of any sutures, it seems safe to assume the extreme flattening of the skull was caused by its natural growth pattern and is not artificial. This too is significant.

Another concern is the absence of the external occipital

protuberance (inion) from its notable position in the center of the occipital, and indeed is represented by a slight fossa (depression) in the surface. (As was mentioned earlier, the same is true for its internal counterpart, which is reduced.) It seems clear this being's neck was attached to its skull much lower than normal and centered under the balance point for both lateral and medial flexion. Even more unusual, the neck seems to have a circumference somewhere in the range of 50 percent of usual neck volume, which presents us with yet another example of the thorough uniqueness of this specimen.

In addition to lacking frontal sinuses, the brow ridges of normal skulls are not evident. Its upper orbits are thin edged rather than rounded. Its zygomatic arches are reduced and lowered from their usual positions. Its mastoid processes are less than normal, as are connective points for the lower face (which would attach to the coronoid process and condylar process of the missing mandible). Based on these observations, its lower face may have been as much as 50 percent less than normal. On the other hand, its inner ears are noticeably larger than normal, again about 50 percent larger. This is also true for the condyles abutting the spinal atlas.

A detached upper right maxilla contains two molars. [*Note from Lloyd Pye: The larger one has since been lost to DNA testing*]. Tooth wear on the molars indicates maturity was reached, yet another set of teeth are present in the maxilla and appear ready to take the place of those mature teeth when and if they are lost or are no longer useful. The question of age at death remains open.

Carbon 14 dating has shown the adult skull to be 900 years old ± 40 years. [*Note from Lloyd Pye: The Starchild skull has also been C-14 tested to 900 years old ± 40 years, supporting the provenance story as given.*] These and other mysteries about this skull await further analysis by other experts wishing to help determine its origin and history.

APPENDIX II

Report on DNA Analysis of Skeletal Remains from Two Skulls
Dr. Jason Eshleman, PhD,
Trace Genetics, Inc.
Richmond, California.

Two sets of remains were received by Trace Genetics and were processed for genetic analyses. The remains consisted of two skulls presented by Mr. Lloyd Pye for DNA analysis.

SAMPLING

Prior to attempts to extract DNA from the remains, the remains were inventoried and taped with a video camera. Video records of the sampling procedure and the initial extraction of all samples were taken and archived by Trace Genetics. Samples were cut from the left parietal of an abnormally shaped skull, identified as the "Starchild" skull, on Feb. 10, 2003. The equipment used to sample was sterilized using a bleach solution prior to use. Sampling was performed in a room not used for any genetic analyses. Fragments weighing a total of 0.8g were cut from the parietal using a rotary cutter with a previously unused blade. The fragments were placed in a sterile conical tube labeled SCS-1 and stored for analysis. A second 0.7g fragment adjacent to the sample retained by Trace Genetics was placed in a sterile conical tube labeled SCS-2 and returned to Mr. Pye.

Two teeth were removed from the maxilla of a skull presented in association with the Starchild skull on February 10, 2003. The right first molar tooth and root weighing 1.7g was removed and labeled SA-1. The tooth and root were placed in a sterile conical tube and retained for genetic analysis by Trace Genetics. A portion of the root was fractured in the process and remained in the maxilla. The right premolar and root (sample labeled SA-2; total weight 1.0g) were also removed from the maxilla, placed in a sterile conical tube. The SA-2 sample was returned to Mr. Pye.

EXTRACTION AND ANALYSIS OF DNA SCS-1

Extraction 1:

A first extraction was performed on a 0.24g fragment of the parietal bone from sample SCS-1. The extraction was performed in a dedicated ancient DNA laboratory beginning on March 7, 2003, and was performed in parallel to an extraction of SA-1 and a reagent blank (negative control). Both surfaces of bone were sanded with a rotary sander to remove any surface contaminants and lacquer preserves present on the outer surfaces of bone. Subsequent to sanding, the bone was exposed to ultraviolet (UV) irradiation (254nm) for 300 seconds per side. The bone surface was then cleaned with bleach (2% sodium hypochlorite), rinsed with sterile EDTA and placed in a fresh 15ml conical tube and immersed in approximately 2ml of 0.5M EDTA. The tube was sealed with parafilm and placed on a rocker.

After 10 days, the tube was opened and 150μl of 0.1M PTB and 20μl of 100mg/ml proteinase K was added to the sample and EDTA. The sample was incubated with agitation overnight at 64° C. DNA was extracted from the digested sample using a 3-step phenol/chloroform extraction method. Two extractions with phenol:chloroform:isoamyl (25:24:1) of equal volume to the digested product were followed by an extraction with an equal volume of chloroform:isoamyl (24:1).

The extracted DNA solution was concentrated by 3 ammonium-acetate precipitation using two volumes of cold 100% filtered ethanol and ½ volume of 5M ammonium acetate. This solution was then stored at −20° C for approximately 4 hours to facilitate precipitation, then centrifuged at high speeds (10,000–12,000rpm) for 15 minutes to pellet the precipitated DNA. The supernatant was discarded and the remaining DNA pellet dried and resuspended in ~300µl sterile ddH2O.

To further purify the DNA and remove additional PCR inhibitors co-extracted with the DNA, the DNA solution was purified using Promega® Wizard PCR Preps DNA Purification Kit as directed by the manufacturer. DNA was eluted from Promega® columns with 100µl sterile ddH2O, the elutant labeled SCSex1 and stored at –20° C.

Attempts to amplify segments of mtDNA from extract SC-Sex1 were performed as described below in AMPLIFICATION METHODS. Single amplifications for fragments containing the diagnostic mutations for Native American haplogroups A, B, C and D did not reveal a known Native American haplogroup; however, the extraction did not amplify consistently.

A single amplification of a fragment of the mtDNA first hypervariable segment (HVSI) between np16210 and np16328 was sequenced using a cycle sequencing procedure with ABI Big-Dye 3.1 chemistry and analyzed on an ABI automated genetic analyzer. The sequence obtained revealed a transition relative to the Cambridge reference sequence at np16273.

This sequence did not match any of the personnel with access to the ancient DNA facilities, nor did it match a sequence obtained from Mr. Pye. Subsequent amplifications of this fragment were not successful and the sequence could not be confirmed. Attempts to amplify fragments of the amelogenin gene located on the X and Y chromosome were uniformly not successful.

Extraction 2:

A second extraction was performed beginning April 21, 2003 on 0.21g of the parietal sample from SCS-1. The extraction was performed as above with the following modifications: 1) The sample was run in parallel with a reagent blank (negative control) but was not processed with any other samples; 2) The bone was exposed to 900 seconds of UV irradiation per side; 3) The bone was completely immersed in 2% sodium hypochlorite for 5 minutes; 4) The sample was left in EDTA with agitation for 22 days prior to digestion with proteinase K; 5) At digestion, ~50µl of Tween-20 was added with 100µl of PTB; and 6) The silica extraction columns (Promega®) were eluted with 80µl of ddH2O and sample labeled SCSe2.

Attempts to amplify mtDNA for fragments containing the diagnostic mutations for Native American haplogroups A, B, C, and D were performed on extract SCSe2. Multiple amplifications indicated the sample possessed an AluI restriction site at np13262 indicative of Native American haplogroup C. Sequence obtained for a fragment of the first hypervariable segment of the mtDNA control region from np16210 to np16367 revealed transitions at np16223, np16298, np16325, and np16327. These mutations are characteristic of haplogroup C in the Americas.

Multiple attempts to amplify a segment of the amelogenin gene were unsuccessful using various amounts of SCSe2 extract as template. 30µl of the original extract was concentrated to a final volume of ~10µl using a microcon YM-30 concentrator. Attempts to amplify this concentrated template were not successful.

Extraction 3:

A third extraction was performed beginning on June 4, 2003, as described above for extraction 2 with the following modifications: 1) The extraction was performed on the remaining 0.40g of bone; 2) The sample was immersed in ~3.5ml of EDTA; 3) To ensure adequate demineralization of the sample, the sample was left immersed in EDTA with agitation for 30 days; 4) The final elution from the silica spin columns (Promega®) was performed twice, each time with 35µl of ddH2O preheated to 65° C.

Attempts to amplify fragments of mtDNA were performed to test for a presence of diagnostic mutations for Native American haplogroups A and C. The sample did not appear to possess the diagnostic HaeIII mutation and np663 indicative of haplogroup A. Multiple amplifications did reveal the presence of the AluI site gain at np13262 indicative of haplogroup C.

A single amplification of a fragment of the amelogenin gene located on the X and Y chromosomes produced a single amplification product 106bp in length. Multiple subsequent 6 amplifications did not reproduce this event, as all subsequent attempts did not produce a PCR product.

EXTRACTION AND ANALYSIS OF DNA SA-1

Extraction 1:

A first extraction was performed on a 0.53g fragment of the molar tooth from sample SA-1 beginning on March 7, 2003, in parallel with an extraction of SCS-1 (above) and a reagent blank (negative control). The extraction was performed in the manner describe above for extraction 1 of SCS-1, save that the tooth's outer surface, which had previous to sampling been firmly rooted in the maxilla, was not sanded and the final elution of the silica spin column (Promega®) was eluted to 100μl and labeled SAex1.

Multiple attempts to amplify segments of mtDNA containing amplifications for fragments of mtDNA containing the diagnostic mutations for Native American haplogroup A revealed a HaeIII restriction site at np663 consistent with known Native American haplogroup A.

Amplifications for fragments containing the diagnostic sites for haplogroups B, C, and D did not show presence of mutations indicative of these haplogroups. A single amplification of a fragment of the mtDNA first hypervariable segment (HVSI) between np16210 and np16327 revealed transitions relative to the Cambridge reference sequence at np16223, np16290 and np16319. These mutations are consistent with Native American haplogroup A.

Multiple amplifications of a fragment of the amelogenin gene on the X and Y chromosomes consistently produced a single band 106bp in length when shown on an electrophoretic gel consistent with DNA from a female.

Extraction 2:

A second extraction was performed beginning April 21, 2003, on 0.42g of the tooth sample from SA-1. The extraction was performed similar to extraction 1 on SA-1 (above) with the following modifications: 1) The sample was not run in parallel to any samples from SCS-1; 2) The sample was immersed in EDTA for 26 days prior to digestion with proteinase-K. The final elution was labeled SAe2 and stored at –20° C.

Multiple amplifications of a mtDNA fragment indicated the presence of a HaeIII restriction site at np663, indicative of Native American haplogroup A. Amplifications of the extraction did not possess the AluI site gain at np16262. Multiple amplifications of a fragment of the amelogenin gene produced a single band when visualized on an electrophoretic gel consistent with DNA from a female.

DISCUSSION

MtDNA from virtually all modern, full-blooded Native Americans belongs to one of five mitochondrial lineages or matrilines (designated haplogroups A, B, C, D, and X) marked by the presence or absence of characteristic restriction sites or by the presence of a nine base pair (9-bp) deletion. Analyses of ancient DNA from Native Americans likewise indicates that these haplogroups constitute virtually all prehistoric Native American individuals as well.

The sample taken from the Starchild skull (SCS-1) has mtDNA consistent with Native American haplogroup C, as revealed through two independent extractions performed on fragments of parietal bone. While a single first extraction did not appear to type similarly, this inconsistent result is likely a product of low-level contamination. This single extraction neither amplified consistently nor was the single sequence of HVSI reproducible.

Contamination could have occurred prior to sampling, introduced in the extraction process, or during PCR amplifications. It is unlikely that any contamination could account for haplogroup C mtDNA, a type not possessed by a researcher with access to the ancient DNA facilities, and reagent blanks did not indicate systematic contamination in extractions.

The sample taken from the associated skull (SA-1) has mtDNA consistent with Native American haplogroup A as determined through both extractions. The sample also appeared be from a female individual as evidenced by repeated amelogenin typing. It is unlikely that contamination could account for the haplogroup A mtDNA as this type is not possessed by any researcher with access to the ancient DNA facilities and the reagent blanks did not indicate systematic contamination in the extractions.

As mtDNA exists in high copy number (upwards of three orders of magnitude relative to a single copy nuclear DNA locus), it can be recovered from prehistoric biological material in sufficient quantities for amplification and analysis using the polymerase chain reaction (PCR). MtDNA is present in haploid condition with inheritance being passed down exclusively through maternal lines. Thus, that the samples analyzed from SCS-1 and SA-1 possessed markedly different mtDNA types excludes a mother-offspring relationship between the two individuals. As it was possible to type and confirm both to known pre-Columbian 9 mtDNA types found in the Americas, both individuals appear to have had Native American mothers.

While it is possible to obtain nuclear DNA from ancient samples, the reduced copy number at a particular nuclear locus relative to mtDNA makes it less likely that a particular extract will contain sufficient DNA for the analysis of a nuclear genetic locus using presently available PCR methods. The ability to amplify nuclear DNA from the SA-1 extractions but not from the SCS-1 extractions could be a product of any of a number of factors.

In ancient DNA analysis, success rates from teeth are generally higher than from bone. Furthermore, there is some indication that X-ray exposure damages and degrades DNA, which may have decreased the quantity and quality of DNA available in the bone prior to extraction. The lone amplification using the amelogenin primers on extract SCSe3 could not be confirmed through additional amplifications and likely indicates a sporadic contamination of a single PCR reaction caused either by a female individual in the laboratory or could have been introduced to laboratory disposables (e.g., pipette tips, PCR reaction tubes).

Such contamination has been noted elsewhere and consequently, any conclusions drawn from the single un-reproduced PCR reaction should not be taken as any reliable indication as to the DNA present in the sample. The presence of reliably typed mtDNA from SCS samples does indicate that mtDNA is present in the bone. The inability to analyze nuclear DNA indicates such DNA is either not present or present in sufficiently low copy number to prevent PCR analysis using methods available at the present time.

AMPLIFICATION METHODS

All reagents used to extract and amplify were first tested to detect any DNA contamination and ancient DNA facilities were cleaned using bleach to remove possible sources of contamination. Further additional contamination controls and precautions are described below.

PCR amplifications of mtDNA were conducted in 25μl volumes using 4μl dNTPs (10mm), 2.5μl 10X PCR buffer (Gibco), 1.3μl BSA (20mg/ml), 0.75μl MgCl2, 0.2μl Platinum Taq DNA polymerase (Gibco) 2 to 6μl of DNA template and sterile ddH20 sufficient to bring reaction volume to 25μl. After an initial 4-minute denaturation step at 94° C to activate the hot start Taq, 40 PCR cycles were performed consisting of a 94° C denaturing step, a 50-55°C annealing step (temperature depending on primers utilized), and a 72° C extension step of 30 seconds each.

A final 3-minute extension at 72° C was added after the last cycle. A portion of the amplification product (~5μl) was run on a 6% polyacrylamide gel together with a size standard ladder, stained with ethidium bromide and photographed under UV light using a digital imager (ISO 2000 imaging system, Alpha Innotech, San Leandro, CA).

To assess the presence or the absence of diagnostic restriction sites, the remaining 20μl were incubated with 10 units of appropriate restriction enzyme overnight at 37°C, then subjected to electrophoresis in the manner earlier described. Amelogenin amplifications were attempted using 1, 3, and 8μl of DNA template in 25μl reaction volumes, adjusting ddH2O amounts to hold concentrations of other reagents.

CONTAMINATION CONTROLS

Ancient DNA is typically highly degraded and survives in much lower copy numbers than modern DNA. Consequently, ancient DNA is highly vulnerable to contamination from modern sources, so specific precautions against contamination were utilized in this study to minimize contamination and, more importantly, to identify contamination when present so that it does not lead to false inferences.

These measures include:
1) Use of dedicated laboratory space, supplies, reagents, and equipment for preparation of ancient DNA samples inside UV-irradiated glove boxes;
2) Use of sterile, disposable labware and clothing whenever possible;
3) Use of separate pre- and post-PCR facilities;
4) Periodic UV irradiation and bleaching of all materials used to help to eliminate any surface contamination;
5) Running negative controls at all stages of the extraction and amplification process to identify the presence of any contaminants;
7) Confirmation of results by multiple amplifications of multiple extractions.

All positive results were confirmed though multiple amplifications of each extraction and multiple extractions performed at different times.

Appendix III

Summary of Inorganic Chemistry Analysis of Starchild Bone

Professor Ken Pye, Director
Kenneth Pye Associates Ltd.
Crowthorne, England

[*Note from Lloyd Pye: Two aspects of chemical analysis can and should be done on the Starchild's bone—organic and inorganic. Professor Ken Pye (no relation to Lloyd Pye) undertook to analyze several inorganic aspects because that is his scientific specialty. While interesting, these are not as revealing as organic analyses would be. Those will have to be carried out at a future time when circumstances permit.*]

The key issues I have addressed in my testing concern: (a) The similarity of the adult female (FEM) and Starchild (SCD) bone samples in terms of chemistry, mineralogy, texture, and radiocarbon age; (b) Whether anything is "unusual" about the Starchild bone in terms of these features; (c) Whether isotopic and trace element chemistry is consistent with one or both samples originating from the area proposed—(as I understand it, the Copper Canyon area of Chihuahua State in Mexico). Further work is still required on these issues, as they are required for the organic chemistry testing, but my preliminary conclusions are that:

(1) There is nothing especially unusual about the inorganic chemistry and mineralogy of the Starchild bone, at least in terms of those aspects examined, which includes major/trace element composition, strontium isotope ratios, lead isotope ratios, oxygen isotope ratio, hydrogen isotope ratio, and

the bone mineralogy. It all falls within the range of values of known human specimens in all respects.

(2) The radiocarbon age determined on collagen extracted from the Starchild bone (900 ± 40 years) is essentially indistinguishable from that previously reported for the human, so they may well be contemporaneous. Owing to the nature of the calibration for the radiocarbon timescale, the radiocarbon age translates to an equivalent of 1 Sigma calibrated age range—Cal AD 1040 to 1190.

(3) In terms of bulk bone mineralogy there are no significant differences between the FEM and SCD bone samples. The sample of female bone analyzed had some particles of quartz-rich dust held within a form of surface patina (the result of exposure to air and/or soil). The quartz peaks are not present in the sample of SCD bone analyzed, but otherwise the traces are indistinguishable.

(4) The SCD and FEM bone samples have very similar strontium isotope and oxygen isotope ratio values. The hydrogen isotope values are also broadly similar although not identical. The lead isotope ratios and several of the chemical element ratios show some significant differences.

(5) Based on the information currently available, these data are consistent with both individuals originating from a relatively high altitude area in the sub-tropics, which would include the mountain/plateau areas of north/central Mexico; however, other possibilities also exist.

The magnitude of the differences between the FEM and SCD bone samples is not large, but does suggest that the Starchild might have had a slightly different diet/water supply, or been brought up in a slightly different area, from the one where the female spent most of her later life.

It should be borne in mind, however, that the Starchild was less than 6 or 7 years old (and maybe only 4) at time of death, whereas the female was an adult and may well have moved to several different areas and changed diet/water supply several times during the course of her life.

Index

Index

Index

Index

Index

Index

Index

279

Index